The staff of the National Hurricane Center receives numerous requests for statistical information on deaths and damages incurred during tropical cyclones affecting the United States. Information about their intensity is also frequently of interest. Estimates of these measures vary in the literature. Our hope is to present the best compilation of currently available estimates. In some instances, data in our lists represent revised estimates based on more complete information received following earlier publications including previous versions of this technical memorandum.

There are other frequently asked questions about hurricanes, such as: What is the average number of hurricanes per year? Which year(s) had the most and least hurricanes? Which hurricane had the longest life? On what date did the earliest and latest hurricane occur? What was the most intense Atlantic hurricane? What was the largest number of hurricanes in existence on the same day? When was the last time a major hurricane[1] or any hurricane impacted a given community? Answers to these and several other questions are provided in Section 3.

---

[1] A major hurricane is a category 3, 4, or 5 hurricane on the Saffir/Simpson Hurricane Scale (see Table 1), and is comparable to a Great Hurricane in some other publications.

Table 1. Saffir/Simpson Hurricane Scale, modifed from Simpson (1974).

Scale Number (Category)	Winds (Mph)	Typical characteristics of hurricanes by category			
		(Millibars)	(Inches)	Surge (Feet)	Damage
1	74-95	> 979	> 28.91	4 to 5	Minimal
2	96-110	965-979	28.50-28.91	6 to 8	Moderate
3	111-130	945-964	27.91-28.47	9 to 12	Extensive
4	131-155	920-944	27.17-27.88	13 to 18	Extreme
5	> 155	< 920	< 27.17	> 18	Catastrophic

2. BACKGROUND AND DEFINITIONS

Many of the statistics in this publication depend directly on the criteria used in preparing another study, "Hurricane Experience Levels of Coastal County Populations-Texas to Maine" [(Jarrell et al. 1992)]. The primary purpose of that study was to demonstrate, county by county, the low hurricane experience level of a large majority of the population. Statistics show that the largest loss of life and property occur in locations experiencing the core of a category 3 or stronger hurricane.

The Saffir/Simpson Hurricane Scale (SSHS, Table 1) provides wind, associated central pressure, and storm surge values. There is not a one-to-one relationship between these elements and it is important to note that the original SSHS category assignment was based on a combination of these elements (Hebert and Taylor 1975). Since about 1990, however, the NHC has assigned the SSHS category by using the maximum one-minute wind speed value only. Thus there is an inconsistency in the HURDAT database (Jarvinen et al. 1984) that will be rectified as the Atlantic best-track reanalysis project is completed (Landsea et al. 2004b). Currently, the SSHS category assignment is based on wind speed from 1851-1914 and 1990-2006 and on a combination of wind, pressure and storm surge from 1915-1989. Heavy rainfall associated with a hurricane is <u>not</u> one of the criteria for categorizing.

Dvorak satellite intensity estimates are often the only estimate of the wind. Available surface wind reports, surface estimates of wind from passive/active microwave satellites, aircraft reconnaissance flight-level winds (from which surface wind speed can be estimated), and dropsonde data occasionally supplement these wind estimates. In post-storm analysis, the central pressure ranges of hurricanes on the SSHS will usually agree fairly well with the wind ranges for each category. On the other hand, the storm surge is strongly dependent on the slope of the continental shelf (shoaling factor). This can <u>change the height of the surge by a factor of two</u> for a given central pressure and/or maximum wind.

The process of assigning a category number to a hurricane in any location is subjective, as is NHC's estimate of a cyclone's impact. It is made on a county-by-county basis. In this study, we use criteria for direct hit as described in the work by Jarrell et al. (1992). Note we are discontinuing the use of the term indirect hit because of the lack of local information that is conveyed in that language.

> <u>Direct Hit</u> - Using "R" as the radius of maximum winds in a hurricane (the distance in miles from the storm's center to the circle of maximum winds around the center), all or parts of coastal counties falling within approximately 2R to the right and R to the left of a storm's track were considered to have received a direct hit. (This assumes an observer at sea looking toward the shore. If there was no landfall, the closest point of approach was used in place of the landfall point). On average, this direct hit zone extended about 50 miles along the coastline (R≈15 miles). Of course, some hurricanes were smaller than this and some, particularly at higher latitudes, were much larger. Cases were judged individually, and many borderline situations had to be resolved.

In this document, the term strike is designated to mean one of two things:

1) During the years 1851-1914 and 1990 to 2006, a hurricane strike is defined as a hurricane that is estimated to have caused sustained hurricane-force winds on the coastline, but does not necessarily make landfall in the area of hurricane-force winds. One example of a hurricane strike is Hurricane Ophelia in 2005, which remained offshore of the North Carolina coast but still brought sustained hurricane-force winds to the coastline.
2) During the years 1915 to 1989, a hurricane strike is defined as a hurricane whose center passes within the direct hit definition area provided above. The best-track reanalysis project is working to change the definition to be strictly defined by the winds, but for now the regional effects catalogued by HURDAT are in a transition period that could last several more years.

Statistics on tropical storm and hurricane activity in the North Atlantic Ocean (which includes the Gulf of Mexico and the Caribbean Sea) can be found in Neumann et al. (1999). A stratification of hurricanes by category which have affected coastal counties of the Gulf of Mexico and North Atlantic Ocean can be found in Jarrell et al. (1992) and also at the NOAA Coastal Services Center (http://hurricane.csc.noaa.gov/hurricanes/index.htm). Additional information about the impact of hurricanes can be found in annual hurricane season articles in Monthly Weather Review, Storm Data and Mariner's Weather Log.

A new feature for this update is including the inland impacts of some hurricanes. These cyclones are indicated with an "I" before the state abbreviation in the HURDAT database and are exclusively used for hurricane impacts that are felt in a state, but not at the coastal areas. One example of this occurrence is Hurricane Dennis (2005). After landfall, Dennis produced category one hurricane winds over inland areas of Alabama, but these effects were not felt along the coast of Alabama. Thus an "I" is added in front of the state designation, to be IAL 1. If a hurricane primarily impacts the coastal areas of a state, inland effects are not listed separately. The goal of this listing is to indicate only the most significant impact of that state. Because of the geography of Florida, any effects in the state are considered coastal.

3. DISCUSSION
Part I

The remainder of this memorandum provides answers to some of the most frequently asked questions about the characteristics and impacts of the tropical cyclones to affect the United States from 1851-2006.

(1) What have been the deadliest tropical cyclones in the United States? Table 2 lists the tropical cyclones that have caused at least 25 deaths on the U.S. mainland 1851-2006. The Galveston Hurricane of 1900 was responsible for at least 8000 deaths and remains #1 on the list. Hurricane Katrina of 2005 killed at least 1500 people and is the third deadliest hurricane to strike the United States. No other landfalling tropical cyclones from 2005 or 2006 made the list. Hurricane Audrey of 1957 has moved up a few places on the list due to an updated list of deaths described in Ross and Goodson (1997). A tropical storm which affected southern California in 1939 and the deadliest Puerto Rico and Virgin Islands hurricanes are listed as addenda to the table.

(2) What have been the costliest tropical cyclones in the United States? Table 3a lists the thirty costliest tropical cyclones to strike the U.S. mainland from 1900-2006. No monetary estimates are available before 1900 and figures are not adjusted for inflation. Hurricane Katrina of 2005 was responsible for at least 81 billion dollars of property damage and is by far the costliest hurricane to ever strike the United States. It is of note that the 2004 and 2005 hurricane seasons produced seven out of the nine costliest systems ever to affect the United States. Table 3b re-orders Table 3a after adjusting to 2006 dollars[2] and adds several other hurricanes. Even after accounting for inflation, the 2004 and 2005 hurricane seasons produced seven out of the thirteen costliest systems ever to strike the United States. Hawaiian, Puerto Rican and Virgin Island tropical cyclones are listed as addenda to Tables 3a and 3b. Table 3b also lists the thirty costliest hurricanes 1900-2006 assuming that a hurricane having the same track, size and intensity as noted in the historical record would strike the area with today's population totals and property-at-risk. Note that the methodology (Pielke and Landsea 1998) which was used to update this technical memorandum for the past two issuances has been changed. See Pielke et al. (2007) for more details.

(3) What have been the most intense hurricanes to strike the United States? Table 4 lists the most intense major hurricanes to strike the U.S. mainland 1851-2006. In this study, hurricanes have been ranked by estimating central pressure at time of landfall. We have used central pressure as a proxy for intensity due to the uncertainties in maximum wind speed estimates for many historical hurricanes. Hurricane Katrina had the third lowest pressure ever noted at landfall, behind the 1935 Florida Keys hurricane and Hurricane Camille in 1969. A total of seven hurricanes from the 2004 and 2005 season had low enough pressures at landfall to be included in the list, five of which placed in the top thirty. Hawaiian, Puerto Rican and Virgin Island hurricanes are listed as addenda to Table 4.

---

[2] Adjusted to 2006 dollars on the basis of U.S. Department of Commerce Implicit Price Deflator for Construction. Available index numbers are rounded to the nearest tenth. This rounding can result in slight changes in the adjusted damage of one hurricane relative to another.

A look at the lists of deadliest and costliest hurricanes reveals several striking facts: (1) Fourteen out of the fifteen deadliest hurricanes were category 3 or higher. (2) Large death totals were primarily a result of the 10 feet or greater rise of the ocean (storm surge) associated with many of these major hurricanes. Katrina of 2005 typifies this point. (3) A large portion of the damage in
four of the twenty costliest tropical cyclones (Table 3a) resulted from inland floods caused by torrential rain. (4) One-third of the deadliest hurricanes were category four or higher. (5) Only six of the deadliest hurricanes occurred during the past twenty five years in contrast to three-quarters of the costliest hurricanes (this drops to sixty percent after adjustment for inflation and about one-quarter after adjustment for inflation, population, and personal wealth).

Katrina provided a grim reminder of what can happen in a hurricane landfall. Sociologists estimate, however, that people only remember the worst effects of a hurricane for about seven years (B. Murrow, personal communication). One of the greatest concerns of the National Weather Service's (NWS) hurricane preparedness officials is that people will _think that no more large loss of life will occur in a hurricane because of our advanced technology and improved hurricane forecasts_. Bill Proenza, spokesman for the NWS hurricane warning service and current Director of NHC, as well as former NHC Directors, have repeatedly emphasized the great danger of a catastrophic loss of life in a future hurricane if proper preparedness plans for vulnerable areas are not formulated, maintained and executed.

The study by Jarrell et al. (1992) used 1990 census data to show that 85% of U.S. coastal residents from Texas to Maine had _never_ experienced a direct hit by a major hurricane. This risk is higher today as an estimated _50 million residents_ have moved to coastal sections during the past twenty-five years. _The experience gained through the landfall of 7 major hurricanes during the past 3 years_ has not lessened an ever-growing concern brought by the continued increase in coastal populations.

Table 2. Mainland U.S. tropical cyclones causing 25 or greater deaths 1851-2006.

RANK	HURRICANE	YEAR	CATEGORY	DEATHS
1	TX (Galveston)	1900	4	8000 [a]
2	FL (SE/Lake Okeechobee)	1928	4	2500 [b]
3	KATRINA (SE LA/MS)	2005	3	1500
4	LA (Cheniere Caminanda)	1893	4	1100-1400 [c]
5	SC/GA (Sea Islands)	1893	3	1000-2000 [d]
6	GA/SC	1881	2	700
7	AUDREY (SW LA/N TX)	1957	4	416 [h]
8	FL (Keys)	1935	5	408
9	LA (Last Island)	1856	4	400 [e]
10	FL (Miami)/MS/AL/Pensacola	1926	4	372
11	LA (Grand Isle)	1909	3	350
12	FL (Keys)/S TX	1919	4	287 [j]
13	LA (New Orleans)	1915	4	275 [e]
13	TX (Galveston)	1915	4	275
15	New England	1938	3	256
15	CAMILLE (MS/SE LA/VA)	1969	5	256
17	DIANE (NE U.S.)	1955	1	184
18	GA, SC, NC	1898	4	179
19	TX	1875	3	176
20	SE FL	1906	3	164
21	TX (Indianola)	1886	4	150
22	MS/AL/Pensacola	1906	2	134
23	FL, GA, SC	1896	3	130
24	AGNES (FL/NE U.S.)	1972	1	122 [f]
25	HAZEL (SC/NC)	1954	4	95
26	BETSY (SE FL/SE LA)	1965	3	75
27	Northeast U.S.	1944	3	64 [g]
28	CAROL (NE U.S.)	1954	3	60
29	FLOYD (Mid Atlantic & NE U.S.)	1999	2	56
30	NC	1883	2	53
31	SE FL/SE LA/MS	1947	4	51 [h,j]
32	NC, SC	1899	3	50 [h,j]
32	GA/SC/NC	1940	2	50
32	DONNA (FL/Eastern U.S.)	1960	4	50 [h]
35	LA	1860	2	47 [h]
36	NC, VA	1879	3	46 [h,j]
36	CARLA (N & Central TX)	1961	4	46
38	TX (Velasco)	1909	3	41
38	ALLISON (SE TX)	2001	TS [k]	41
40	Mid-Atlantic	1889	none [l]	40 [h,j]
40	TX (Freeport)	1932	4	40
40	S TX	1933	3	40

RANK	HURRICANE	YEAR	CATEGORY	DEATHS
43	HILDA (LA)	1964	3	38
44	SW LA	1918	3	34
45	SW FL	1910	3	30
45	ALBERTO (NW FL, GA, AL)	1994	TS [k]	30
47	SC, FL	1893	3	28 [m]
48	New England	1878	2	27 [h,n]
48	Texas	1886	2	27 [h]
50	FRAN (NC)	1996	3	26
51	LA	1926	3	25
51	CONNIE (NC)	1955	3	25
51	IVAN (NW FL, AL)	2004	3	25
ADDENDUM (Not Atlantic/Gulf Coast)				
2	Puerto Rico	1899	3	3369 [i]
6	P.R., USVI	1867	3	811 [f,j]
6	Puerto Rico	1852	1	800 [f,o]
12	Puerto Rico (San Felipe)	1928	5	312
17	USVI, Puerto Rico	1932	2	225
25	DONNA (St. Thomas, VI)	1960	4	107
25	Puerto Rico	1888	1	100 [h]
38	Southern California	1939	TS [k]	45
38	ELO/SE (Puerto Rico)	1975	TS [k]	44
48	USVI	1871	3	27 [h]

Notes:
a Could be as high as 12,000
b Could be as high as 3000
c Total including offshore losses near 2000
d August
e Total including offshore losses is 600
f No more than
g Total including offshore losses is 390
h At least
i Puerto Rico 1899 and NC, SC 1899 are the same storm
j Could include some offshore losses
k Only of Tropical Storm intensity.
l Remained offshore
m Mid-October
n Four deaths at shoreline or just offshore
o Possibly a total from two hurricanes

Table 3a. The thirty costliest mainland United States tropical cyclones, 1900-2006, (not adjusted for inflation).

RANK	HURRICANE	YEAR	CATEGORY	DAMAGE (U.S.)
1	KATRINA (SE FL, SE LA, MS)	2005	3	$81,000,000,000
2	ANDREW (SE FL/SE LA)	1992	5	26,500,000,000
3	WILMA (S FL)	2005	3	20,600,000,000
4	CHARLEY (SW FL)	2004	4	15,000,000,000
5	IVAN (AL/NW FL)	2004	3	14,200,000,000
6	RITA (SW LA, N TX)	2005	3	11,300,000,000
7	FRANCES (FL)	2004	2	8,900,000,000
8	HUGO (SC)	1989	4	7,000,000,000
9	JEANNE (FL)	2004	3	6,900,000,000
10	ALLISON (N TX)	2001	TS @	5,000,000,000
11	FLOYD (Mid-Atlantic & NE U.S.)	1999	2	4,500,000,000
12	ISABEL (Mid-Atlantic)	2003	2	3,370,000,000
13	FRAN (NC)	1996	3	3,200,000,000
14	OPAL (NW FL/AL)	1995	3	3,000,000,000
15	FREDERIC (AL/MS)	1979	3	2,300,000,000
16	DENNIS (NW FL)	2005	3	2,230,000,000
17	AGNES (FL/NE U.S.)	1972	1	2,100,000,000
18	ALICIA (N TX)	1983	3	2,000,000,000
19	BOB (NC, NE U.S)	1991	2	1,500,000,000
19	JUAN (LA)	1985	1	1,500,000,000
21	CAMILLE (MS/SE LA/VA)	1969	5	1,420,700,000
22	BETSY (SE FL/SE LA)	1965	3	1,420,500,000
23	ELENA (MS/AL/NW FL)	1985	3	1,250,000,000
24	GEORGES (FL Keys, MS, AL)	1998	2	1,155,000,000
25	GLORIA (Eastern U.S.)	1985	3	900,000,000
26	LILI (SC LA)	2002	1	860,000,000
27	DIANE (NE U.S.)	1955	1	831,700,000
28	BONNIE (NC,VA)	1998	2	720,000,000
29	ERIN (NW FL)	1998	2	700,000,000
30	ALLISON (N TX)	1989	TS @	500,000,000
30	ALBERTO (NW FL,GA,AL)	1994	TS @	500,000,000
30	FRANCES (TX)	1998	TS @	500,000,000
30	ERNESTO (FL,NC,VA)	2006	TS @	500,000,000

ADDENDUM (Rank is independent of other events in group)

19	GEORGES (USVI,PR)	1998	3	1,800,000,000
19	INIKI (Kauai, HI)	1992	3	1,800,000,000
19	MARILYN (USVI, PR)	1995	2	1,500,000,000
25	HUGO (USVI, PR)	1989	4	1,000,000,000
30	HORTENSE (PR)	1996	1	500,000,000

Notes:

@ Only of Tropical Storm intensity

Table 3b. The thirty costliest mainland United States tropical cyclones, 1900-2006.

	Ranked Using 2006 Deflator*					Ranked Using 2006 Inflation, Population and Wealth Normalization[L]			
RANK	HURRICANE	YEAR	Category	Damage (Millions)*		RANK	HURRICANE	YEAR Category	Damage (Millions)[L]
1	KATRINA (LA/MS/FL)	2005	3	$84,645		1	SE Florida/Alabama	1926 4	$164,839
2	ANDREW (SE FL/SE LA)	1992	5	48,058		2	N Texas (Galveston)	1900 4	104,330
3	WILMA (SW/SE FL)	2005	3	21,527		3	KATRINA (SE LA, MS, AL)	2005 3	85,050
4	CHARLEY (SW FL)	2004	4	16,322		4	N Texas (Galveston)	1915 4	71,397
5	IVAN (NW FL/AL)	2004	3	15,451		5	ANDREW (SE FL/LA)	1992 5	58,555
6	HUGO (SC)	1989	4	13,480		6	New England	1938 3	41,122
7	AGNES (FL/NE U.S.)	1972	1	12,424		7	SW Florida	1944 3	40,621
8	BETSY (SE FL/SE LA)	1965	3	11,883		8	SE Florida/Lake Okeechobee	1928 4	35,298
9	RITA (LA/TX/FL)	2005	3	11,808		9	DONNA (FL/Eastern U.S.)	1960 4	28,159
10	CAMILLE (MS/SE LA/VA)	1969	5	9,781		10	CAMILLE (MS/LA/VA)	1969 5	22,286
11	FRANCES (SE FL)	2004	2	9,684		11	WILMA (S FL)	2005 3	21,630
12	DIANE (NE U.S.)	1955	1	7,700		12	BETSY (SE FL/LA)	1965 3	18,749
13	JEANNE (SE FL)	2004	3	7,508		13	DIANE (NE U.S.)	1955 1	18,073
14	FREDERIC (AL/MS)	1979	3	6,922		14	AGNES (NW FL, NE U.S.)	1972 1	18,052
15	New England	1938	3	6,571		15	HAZEL (SC/NC)	1954 4	17,339
16	ALLISON (N TX)	2001	TS	6,414		16	CHARLEY (SW FL)	2004 4	17,135
17	FLOYD (Mid Atlc & NE U.S.)	1999	2	6,342		17	CAROL (NE U.S.)	1954 3	16,940
18	NE U.S.	1944	3	5,927		18	IVAN (NW FL, AL)	2004 3	16,247
19	FRAN (NC)	1996	3	4,979		19	HUGO (SC)	1989 4	16,088
20	ALICIA (N TX)	1983	3	4,825		20	SE Florida	1949 3	15,398
21	OPAL (NW FL/AL)	1995	3	4,758		21	CARLA (N & Central TX)	1961 4	14,920
22	CAROL (NE U.S.)	1954	3	4,345		22	SE Florida/Louisiana/Alabama	1947 4	14,406
23	ISABEL (NC/VA)	2003	2	3,985		23	NE U.S.	1944 3	13,881
24	JUAN (LA)	1985	1	3,417		24	S Texas	1919 4	13,847
25	DONNA (FL/Eastern U.S.)	1960	4	3,345		25	SE Florida	1945 3	12,956
26	CELIA (S TX)	1970	3	3,038		26	RITA (SW LA/N TX)	2005 3	11,865
27	BOB (NC, NE U.S)	1991	2	2,853		27	FREDERIC (AL/MS)	1979 3	10,781
28	ELENA (MS/AL/NW FL)	1985	3	2,848		28	FRANCES (SE FL)	2004 2	10,168
29	CARLA (N & Central TX)	1961	4	2,604		29	NC/VA	1933 2	8,603
30	DENNIS (NW FL)	2005	3	2,330		30	DORA (NE FL)	1964 2	8,066
ADDENDUM						notes			
30	INIKI (Kauai, HI)	1992	4	2,563		*	2006 $ based on U.S. DOC Implicit Price Deflator for Construction.		
30+	GEORGES (USVI,PR)	1998	3	2,276		1	Damage estimate in 1915 reference is considered too high		
30+	MARILYN (USVI,E. PR)	1995	2	1,900		[L]	'Normalization reflects inflation, changes in personal wealth and coastal		
30+	HUGO (USVI, PR)	1989	4	1,502			county population to 2005, (Pielke et al. 2007) then including an		
30+	San Felipe (PR)	1928	5	1,424			estimate to 2006 dollars by increasing totals by 5%.		

Table 4. The most intense mainland United States hurricanes ranked by pressure, 1851-2006 (includes only major hurricanes at their most intense landfall).

RANK	HURRICANE	YEAR	CATEGORY (at landfall)	MINIMUM PRESSURE Mill bars	MINIMUM PRESSURE Inches
1	FL (Keys)	1935	5	892	26.35
2	CAMILLE (MS/SE LA/VA)	1969	5	909	26.84
3	KATRINA (SE LA, MS)	2005	3	920	27.17
4	ANDREW (SE FL/SE LA)	1992	5	922	27.23
5	TX (Indianola)	1886	4	925	27.31
6	FL (Keys)/S TX	1919	4	927	27.37
7	FL (Lake Okeechobee)	1928	4	929	27.43
8	DONNA (FL/Eastern U.S.)	1960	4	930	27.46
9	LA (New Orleans)	1915	4	931	27.49
9	CARLA (N & Central TX)	1961	4	931	27.49
11	LA (Last Island)	1856	4	934	27.58
11	HUGO (SC)	1989	4	934	27.58
13	FL (Miami)/MS/AL/Pensacola	1926	4	935	27.61
14	TX (Galveston)	1900	4	936	27.64
15	RITA (SW LA/N TX)	2005	3	937	27.67
16	GA/FL (Brunswick)	1898	4	938	27.70
16	HAZEL (SC/NC)	1954	4	938	27.70
18	SE FL/SE LA/MS	1947	4	940	27.76
19	N TX	1932	4	941	27.79
19	CHARLEY (SW FL)	2004	4	941	27.79
21	GLORIA (Eastern U.S.)	1985	3	942	27.82
21	OPAL (NW FL/AL)	1995	3	942	27.82
23	FL (Central)	1888	3	945	27.91
23	E NC	1899	3	945	27.91
23	AUDREY (SW LA/N TX)	1957	4	945	27.91
23	TX (Galveston)	1915	4	945	27.91
23	CELIA (S TX)	1970	3	945	27.91
23	ALLEN (S TX)	1980	3	945	27.91
29	New England	1938	3	946	27.94
29	FREDERIC (AL/MS)	1979	3	946	27.94
29	IVAN (AL, NW FL)	2004	3	946	27.94
29	DENNIS (NW FL)	2005	3	946	27.94
33	NE U.S.	1944	3	947	27.97
33	SC/NC	1906	3	947	27.97
35	LA (Chenier Caminanda)	1893	4	948	27.99
35	BETSY (SE FL/SE LA)	1965	3	948	27.99
35	SE FL/NW FL	1929	3	948	27.99
35	SE FL	1933	3	948	27.99
35	S TX	1916	3	948	27.99
35	MS/AL	1916	3	948	27.99
41	NW FL	1882	3	949	28.02
41	DIANA (NC)	1984	3	949	28.02
41	S TX	1933	3	949	28.02
44	GA/SC	1854	3	950	28.05
44	LA/MS	1855	3	950	28.05
44	LA/MS/AL	1860	3	950	28.05
44	LA	1879	3	950	28.05
44	BEULAH (S TX)	1967	3	950	28.05
44	HILDA (Central LA)	1964	3	950	28.05
44	GRACIE (SC)	1959	3	950	28.05
44	TX (Central)	1942	3	950	28.05
44	JEANNE (FL)	2004	3	950	28.05
44	WILMA (S FL)	2005	3	950	28.05
54	SE FL	1945	3	951	28.08
54	BRET (S TX)	1999	3	951	28.11
56	LA (Grand Isle)	1909	3	952	28.11
56	FL (Tampa Bay)	1921	3	952	28.11
56	CARMEN (Central LA)	1974	3	952	28.11
59	SC/NC	1885	3	953	28.14
59	S FL	1906	3	953	28.14
61	GA/SC	1893	3	954	28.17
61	EDNA (New England)	1954	3	954	28.17
61	SE FL	1949	3	954	28.17
61	FRAN (NC)	1996	3	954	28.17
65	Central FL	1871	3	955	28.20
65	LA/TX	1886	3	955	28.20
65	SC/NC	1893	3	955	28.20
65	NW FL	1894	3	955	28.20
65	ELOISE (NW FL)	1975	3	955	28.20
65	KING (SE FL)	1950	3	955	28.20
65	Central LA	1926	3	955	28.20
65	SW LA	1918	3	955	28.20

RANK	ADDENDUM	YEAR	CATEGORY (at landfall)	Mill bars	Inches
4	DAVID (S of PR)	1979	4	924	27.29
8	San Felipe (PR)	1928	5	931	27.49
16	HUGO (USVI & PR)	1989	4	940	27.76
41	INIKI (KAUAI, HI)	1992	3	950	27.91
60	DOT (KAUAI, HI)	1959	3	955	28.11

Table 5 summarizes the hurricane strikes on the U. S. mainland since 1851. The data indicate that an average of about 2 major hurricanes every 3 years made landfall somewhere along the U.S. Gulf or Atlantic coast. (All categories combined average about 5 hurricanes every 3 years.) Note that not all areas of the U.S. were settled before 1900 and there could be substantial gaps in landfall data coverage, especially in South Florida. For more details see Landsea et al. (2004b).

Table 5. Hurricane strikes on the mainland United States (1851-2006).

Category	Strikes
5	3
4	18
3	75
2	73
1	110
TOTAL	279
MAJOR	96

Major hurricanes are categories 3, 4 & 5.

Table 6, which lists hurricanes by decades since 1851, shows that during the forty year period 1961-2000 both the number and intensity of landfalling U.S. hurricanes decreased sharply. Based on 1901-1960 statistics, the expected number of hurricanes and major hurricanes during the period 1961-2000 was 75 and 28, respectively. But, in fact, only 55 (or 74%) of the expected number of hurricanes struck the U.S. with only 19 major hurricanes or 68% of that expected number. However, landfall activity during the 2000's has picked up significantly, and is now near the frequency seen in the very active 1940's. These increased landfalls are very different than the late 1990's, which showed average landfall frequencies despite having generally active seasons.

Despite the increase in overall activity, the United States hasn't seen a significant resurgence of exceptionally strong hurricane landfalls. During the past 35 years, the United States has experienced three Category 4 or stronger hurricanes: Charley in 2004, Andrew of 1992 and Hugo of 1989. However, on the average, a category 4 or stronger hurricane strikes the United States about once every 7 years. This suggests we have seen fewer exceptionally strong hurricanes than an expected 35-year average of about 5. Fewer hurricanes, however, do not necessarily mean a lesser threat of disaster. Records for the most intense U.S. hurricane in 1935, and

Table 6. Number of hurricanes by category to strike the mainland U.S. each decade. (Updated from Blake et al., 2005)

DECADE	1	2	3	4	5	ALL 1,2,3,4,5	Major 3,4,5
1851-1860	7	5	5	1	0	18	6
1861-1870	8	6	1	0	0	15	1
1871-1880	7	6	7	0	0	20	7
1881-1890	8	9	4	1	0	22	5
1891-1900	8	5	5	3	0	21	8
1901-1910	10	4	4	0	0	18	4
1911-1920	10	4	4	3	0	21	7
1921-1930	5	3	3	2	0	13	5
1931-1940	4	7	6	1	1	19	8
1941-1950	8	6	9	1	0	24	10
1951-1960	8	1	6	3	0	18	9
1961-1970	3	5	4	1	1	14	6
1971-1980	6	2	4	0	0	12	4
1981-1990	9	2	3	1	0	15	4
1991-2000	3	6	4	0	1	14	5
2001-2006	6	2	6	1	0	15	7
1851-2006	110	73	75	18	3	279	96
Average per decade	7.1	4.7	4.8	1.2	0.2	17.9	6.2

Note: Only the highest category to affect the U.S. has been used

the second costliest, Andrew in 1992, occurred in years which had much below-average hurricane activity. As occurred in Katrina, a large death toll in a U.S. hurricane is still possible, especially in such vulnerable areas as Houston, New York City, Tampa, and the Florida Keys. The decreased death totals in recent years, outside of 2005, is partly the result of relatively few major hurricanes striking the most vulnerable areas.

Continued coastal growth and inflation will almost certainly result in every future major landfalling hurricane (and even weaker hurricanes and tropical storms) replacing one of the current costliest hurricanes. For example, four out of six hurricane landfalls of 2005 made the top 30 list. If warnings are heeded and preparedness plans developed, the death toll can be minimized. In the absence of a change of attitude, policy, or laws governing building practices (codes and location) near the ocean, however, large property losses are inevitable.

Part II

This section answers some frequently asked questions about tropical storm and hurricane activity.

(1) What is the average number of hurricanes per year? Table 7 gives the average number of tropical cyclones which reached tropical storm, hurricane and major hurricane strength during selected time periods. A total of eleven tropical systems reaching storm strength with six of these becoming hurricanes and two attaining major hurricane status are the best averages to use based on the period of geostationary satellite surveillance.

Table 7. Average number of tropical cyclones* which reached storm, hurricane and major hurricane status. Updated from Blake et al. (2005).

PERIOD	Number of Years	Average number of Tropical Storms	Average number of Hurricanes	Average number of Major Hurricanes
1851 - 2006	156	8.7	5.3	1.8
1944[#] - 2006	63	10.6	6.1	2.7
1957 - 2006	50	10.7	6.0	2.4
1966[$] - 2006	41	11.1	6.2	2.3
1977 - 2006	30	11.4	6.3	2.5
1987 - 2006	20	12.6	6.8	2.9
1997 - 2006	10	14.5	7.8	3.6

*Includes subtropical storms after 1967
[#]Start of aircraft reconnaissance
[$]Start of geostationary satellite coverage

(2) What year(s) have had the most/least hurricanes and landfalls?
Table 8a shows the years of maximum and minimum tropical storm, hurricane, and major hurricane activity for the Atlantic hurricane basin. Minimum tropical cyclone activity prior to the satellite surveillance era is uncertain and likely to be underrepresented. Activity during 2005 was far above the previous records for the most number of tropical storms and hurricanes, but 1950 is still the record-holder for the maximum number of major hurricanes. The two year period of 2004-2005 was one of the most active ever seen in the Atlantic basin, setting records for most number of tropical storms and hurricanes in a two year period and tying the record (13) for the most number of major hurricanes set in 1950-1951. It is also of note that seven out of the last twelve years have experienced fourteen or more tropical storms.

Table 8a. Years of maximum and minimum tropical storm, hurricane, and major hurricane activity in the Atlantic basin 1851-2006. Updated from Neumann et al. (1999).

MAXIMUM ACTIVITY					
TROPICAL STORMS[1]		HURRICANES		MAJOR HURRICANES	
Number	Years	Number	Years	Number	Years
28	2005	15	2005	8	1950
21	1933	12	1969	7	1961, 2005
19	1887,1995	11	1887,1916,1950,	6	1916,1926,1955,
18	1969		1995		1964,1996,2004
16	1936,2003	10	1870,1878,1886,	5	1893,1933,1951,
15	2000,2001,2004		1893,1933,1998		1958,1969,
14	1916,1953,1990	9	1880,1955,1980,		1995,1999
	1998		1996,2001,2004	4	occurred in 7 yrs
MINIMUM ACTIVITY*					
TROPICAL STORMS[1]		HURRICANES		MAJOR HURRICANES	
Number	Years	Number	Years	Number	Years
1	1914	0	1907,1914	0	occurred in 33 yrs last in 1994
2	1925,1930	1	1905,1919,1925		
3	1917,1919,1929	2	1851,1854,1890,	1	occurred in 48 yrs last in 1997
4	1854,1857,1868, 1883,1884,1890, 1911,1913,1920, 1983		1895,1917,1922, 1930,1931,1982		

Notes
[1] Includes subtropical storms after 1967.
*likely underrepresented before reconnaissance in 1944

Table 8b lists the years of maximum U.S. hurricane and major hurricane strikes. 2005 set the record for the most U.S. major hurricane strikes since 1851 and tied for second-most hurricane strikes. 2004-2005 produced twelve U.S. hurricane strikes, eclipsing the previous record of eleven hurricane strikes in consecutive years, set in 1886-1887. 2006 did not have a hurricane strike, and the only times that the United States has gone as long as two years without a hurricane strike are 1862-64, 1930-31, 1981-82 and 2000-01. Note there is considerable uncertainty before 1900 because significant areas of the Gulf and Southeast Atlantic coasts were unpopulated. Three or four hurricanes have struck the United States in one year a total of 37 times. Multiple U.S. major hurricane strikes in one year are somewhat rare, occurring on average about once every decade.

Table 8b. Years of maximum United States hurricane and major hurricane strikes 1851-2006.

MAXIMUM U.S. ACTIVITY			
HURRICANE STRIKES		MAJOR HURRICANE STRIKES	
Number	Years	Number	Years
7	1886	4	2005
6	1916,1985,2004,2005	3	1893,1909,1933,
5	1893,1909,1933		1954,2004
4	1869,1880,1887,1888 ,1906,1964	2	1879,1886,1915, 1916,1926,1944,
3	31 years have exactly 3 strikes		1950,1955,1985

(3) When did the earliest and latest hurricanes occur? The hurricane season is defined as June 1 through November 30. The earliest observed hurricane in a year in the Atlantic was on March 7, 1908, while the latest observed hurricane was on December 31, 1954, the second "Alice" of that year which persisted as a hurricane until January 5, 1955. Zeta of 2005 was the second latest tropical cyclone to form, just six hours ahead of Alice 1954. The earliest hurricane to strike the United States was Alma which struck northwest Florida on June 9, 1966. The latest hurricane to strike the United States was late on November 30, 1925 near Tampa, Florida.

(4) What were the longest-lived and shortest-lived hurricanes? The third system of 1899 holds the record for most days as a tropical storm (28) and major hurricane (11.5), while Ginger in 1971 holds the record for the most days as a hurricane (20). There have been many tropical cyclones which remained at hurricane intensity for 12 hours or less, most recently Ernesto of 2006.

(5) What was the hurricane with the lowest central pressure in the Atlantic basin? Wilma in 2005 had an estimated pressure of 882 millibars in the northwestern Caribbean Sea, breaking the record of 888 millibars, previously held by Gilbert of 1988. The 1935 Labor Day hurricane in the Florida Keys had the lowest central pressure in any hurricane to strike the United States since 1851, with a pressure of 892 millibars.

(6) What were the strongest and weakest hurricanes in terms of maximum sustained winds? The Atlantic re-analysis project is undergoing an extensive overhaul of the best track database at this time. Right now, reliable wind estimates are only available for the years 1851-1914 and from about 1990-2006 using modern techniques. After this project is complete, NHC will publish a list of the strongest hurricanes in terms of winds. Numerous hurricanes have reached only the minimum wind speed near 74 miles per hour and made landfall in the United States, most recently Cindy of 2005.

(7) What was the largest number of hurricanes in the Atlantic Ocean at the same time? Four hurricanes occurred simultaneously on two occasions. The first occasion was August 22, 1893, and one of these eventually killed 1,000-2,000 people in Georgia-South Carolina. The second occurrence was September 25, 1998, when Georges, Ivan, Jeanne and Karl persisted into September 27, 1998 as hurricanes. Georges ended up taking the lives of thousands in Haiti. In 1971 from September 10 to 12, there were five tropical cyclones at the same time; however, while most of these ultimately achieved hurricane intensity, there were never more than two hurricanes at any one time.

(8) How many hurricanes have there been in each month? Table 9a, adapted from Neumann et al. (1999), shows the total and average number of tropical storms, and those which became hurricanes and major hurricanes, by month, for the period 1851-2006. Table 9b displays the same statistics from 1966-2006 corresponding to the geostationary satellite era. Table 9a also adds the monthly total and average number of hurricanes to strike the U. S. since 1851.

Table 9a. Tropical storms, hurricanes and major hurricanes in the Atlantic, Caribbean and Gulf of Mexico by month of formation, 1851-2006 [adapted from Neumann et al. (1999)], and for hurricanes striking the U.S. mainland 1851-2006 [updated from Blake et al. (2005)].

MONTH	TROPICAL STORMS[1] Total	Average	HURRICANES Total	Average	MAJOR HURRICANES Total	Average	U.S. HURRICANES Total	Average
JANUARY-APRIL	5	*	1	*	0	0.00	0	0.00
MAY	18	0.1	4	*	1	*	0	0.00
JUNE	79	0.5	28	0.2	3	*	19	0.12
JULY	101	0.6	50	0.3	8	0.05	25	0.16
AUGUST	344	2.2	217	1.4	77	0.50	74	0.48
SEPTEMBER	457	2.9	318	2.0	140	0.92	105	0.69
OCTOBER	280	1.8	158	1.0	53	0.35	51	0.33
NOVEMBER	61	0.4	38	0.2	6	*	5	*
DECEMBER	9	0.1	5	*	0	0.00	0	0.00
YEAR	1354	8.7	819	5.3	288	1.85	279	1.79

[1] Includes subtropical storms after 1967. See Neumann et al. (1999) for details.
* Less than 0.05.

Table 9b. Tropical storms, hurricanes and major hurricanes in the Atlantic, Caribbean and Gulf of Mexico by month of formation, 1966-2006.

MONTH	TROPICAL STORMS[1] Total	Average	HURRICANES Total	Average	MAJOR HURRICANES Total	Average
JANUARY-APRIL	3	0.1	0	*	0	*
MAY	4	0.1	1	*	0	*
JUNE	25	0.6	7	0.2	1	*
JULY	44	1.1	17	0.4	3	0.07
AUGUST	126	3.1	67	1.6	24	0.59
SEPTEMBER	149	3.6	101	2.5	50	1.22
OCTOBER	74	1.8	40	1.0	13	0.32
NOVEMBER	25	0.6	18	0.4	3	0.07
DECEMBER	5	0.1	2	*	0	*
YEAR	455	11.1	253	6.2	94	2.3

[1] Includes subtropical storms after 1967. See Neumann et al. (1999) for details.
* Less than 0.05.

(9) How many direct hits by hurricanes of various categories have affected each state?
Table 10, updated from Blake et al. (2005), shows the number of hurricanes affecting the United States and individual states, i.e., direct hits. Note that the inland information contained in Table 10 does not reflect all storms to affect inland areas. The inland designation is only used for those hurricanes that exclusively struck inland portions of a state (not at the coast). The table shows that, on the average, close to seven hurricanes every four years (~1.8 per year) strike the United States, while about two major hurricanes cross the U.S. coast every three years. Other noteworthy facts, updated from Blake et al. (2005), are: 1.) Forty percent of all U.S. hurricanes and major hurricanes hit Florida; 2.) Eighty-three percent of category 4 or higher hurricane strikes have hit either Florida or Texas; 3.) Sixty percent of all hurricanes affecting Georgia actually come from the south or southwest across northwestern Florida, though these hurricanes from the Gulf of Mexico are much weaker by the time they reach Georgia than the those that come from the Atlantic Ocean. It should be noted that both Florida and Texas have extensive coastlines, which is reflected in the number of occurrences.

Table 10. Hurricane strikes 1851-2006 on the mainland U.S. coastline, and for individual states, including inland areas if effects were only inland portions of the state, by Saffir/Simpson category. Updated from Blake et al. (2005).

AREA	1	2	CATEGORY NUMBER 3	4	5	ALL	MAJOR HURRICANES
U.S. (Texas to Maine)	110	73	75	18	3	279	96
Texas	23	18	12	7	0	60	19
(North)	12	7	3	4	0	26	7
(Central)	7	5	2	2	0	16	4
(South)	7	7	7	1	0	22	8
Louisiana	18	14	15	4	1	52	20
Mississippi	2	5	8	0	1	16	9
Alabama	16	4	6	0	0	26	6
(Inland only)	6	0	0	0	0	6	0
Florida	43	33	29	6	2	113	37
(Northwest)	26	17	14	0	0	57	14
(Northeast)	12	8	1	0	0	21	1
(Southwest)	18	10	8	4	1	41	13
(Southeast)	13	13	11	3	1	41	15
Georgia	15	5	2	1	0	23	3
(Inland only)	9	0	0	0	0	9	0
South Carolina	18	6	4	2	0	30	6
North Carolina	24	14	11	1	0	50	12
(Inland only)	3	0	0	0	0	3	0
Virginia	7	2	1	0	0	10	1
(Inland only)	2	0	0	0	0	2	0
Maryland	1	1	0	0	0	2	0
Delaware	2	0	0	0	0	2	0
New Jersey	2	0	0	0	0	2	0
Pennsylvania (Inland)	1	0	0	0	0	1	0
New York	6	1	5	0	0	12	5
Connecticut	5	3	3	0	0	11	3
Rhode Island	3	2	4	0	0	9	4
Massachusetts	6	2	3	0	0	11	3
New Hampshire	1	1	0	0	0	2	0
Maine	5	1	0	0	0	6	0

Notes:

*State totals will not equal U.S. totals, and Texas or Florida totals will not necessarily equal sum of sectional totals. Regional definitions are found in Appendix A
*Gulf Coast state totals will likely be underrepresented because of lack of coastal population before 1900

(10) When are the major hurricanes likely to strike given areas? Table 11 shows the incidence of major hurricanes by months for the U.S. mainland and individual states. September has about many major hurricane landfalls as October and August combined. The northern Gulf Coast from Texas to Northwest Florida is the prime target for pre-August major hurricanes. The threat of major hurricanes increases from west to east during August with major hurricanes favoring the U.S. East Coast by late September. Most major October hurricanes occur in southern Florida.

Table 11. Incidence of major hurricane direct hits on the U.S. mainland and individual states, 1851-2006, by month. Updated from Blake et al. (2005).

AREA	JUNE	JULY	AUG.	SEPT.	OCT.	ALL
U.S. (Texas to Maine)	2	4	30	44	16	96
Texas	1	1	10	7		19
c (North)	1	1	3	2		7
b (Central)			2	2		4
a (South)			5	3		8
Louisiana	2		7	8	3	20
Mississippi		1	4	4		9
Alabama		1	1	4		6
Florida		2	6	19	10	37
a (Northwest)		2	1	7	3	13
d (Northeast)				1		1
b (Southwest)			2	5	6	13
c (Southeast)			4	8	3	15
Georgia			1	1	1	3
South Carolina			2	2	2	6
North Carolina			4	8	1	13
Virginia				1		1
Maryland						0
Delaware						0
New Jersey						0
Pennsylvania						0
New York			1	4		5
Connecticut			1	2		3
Rhode Island			1	3		4
Massachusetts				3		3
New Hampshire						0
Maine						0

Notes: *State totals do not equal U.S. totals and Texas or Florida totals do not necessarily equal the sum of sectional entries.
*Regional definitions are found in Appendix A.
*Gulf Coast states will likely be underrepresented because of a lack of coastal population before 1900.

(11) How long has it been since a hurricane or a major hurricane hit a given community?
A chronological list of all known hurricanes to strike the United States 1851 through 2006 including month, states affected by category of hurricane, and minimum sea level pressure at landfall can be found in Appendix A, updated from Blake et al. (2005). Table 12 summarizes the occurrence of the last hurricane and major hurricane to directly hit the counties or parishes where most populated coastal communities are located from Brownsville, Texas to Eastport, Maine. An estimated return period of these hurricanes is also listed, which is computed from HURISK (Neumann 1987). In order to obtain the same type of information listed in Table 12 for the remaining coastal communities, the reader is again referred to Jarrell et al. (1992) or the NOAA Coastal Services Center (http://hurricane.csc.noaa.gov/hurricanes/index.htm). There are many illustrative examples of the uncertainty of when a hurricane might strike a given locality. After nearly 70 years without a direct hit, Pensacola, Florida was struck in a period of 11 years by Hurricane Erin and major Hurricane Opal in 1995, major Hurricane Ivan in 2004 and major Hurricane Dennis in 2005. Miami, which expects a major hurricane every nine years, on average, has been struck only once since 1950 (in 1992). Tampa has not experienced a major hurricane for 86 years. Many locations along the Gulf and Atlantic coasts have not experienced a major hurricane during the period 1851-2006 (see Table 12), despite the recent upswing in overall activity.

(12) What is the total United States damage (before and after adjustment for inflation) and death toll for each year since 1900? Table 13a summarizes this information. Table 13b ranks the top 30 years by deaths, unadjusted damage, adjusted damage and normalized damage. In most years the death and damage totals are the result of a single, major hurricane. Gentry (1966) gives damages adjusted to 1957-59 costs as a base for the period 1915-1965. For the most part, death and damage totals for the period 1915-1965 were taken from Gentry's paper, and for the remaining years from Monthly Weather Review. Adjusted damages were converted to 2006 dollars by the factors used in Table 3a.

(13) What are the deadliest and costliest hurricanes to affect Hawaii, Puerto Rico and the U.S. Virgin Islands since 1900? Table 14, provided by Hans Rosendal and Raphael Mojica of the National Weather Service Forecast Offices in Honolulu and San Juan, respectively, summarizes this information. Iniki in 1992 is the deadliest and costliest hurricane to affect Hawaii while Georges of 1998 is the costliest hurricane to affect Puerto Rico. The notorious San Felipe hurricane of 1928 was the deadliest hurricane in Puerto Rico since 1900.

Table 12. Last direct hit and mean return period (Neumann 1987) of a major hurricane or hurricane by county/parish within 75 n mi for certain populated coastal communities. Category in parenthesis.

State	City (County/Parish)	MAJOR HURRICANE Return Period	MAJOR HURRICANE Last Direct Hit By County	HURRICANE Return Period	HURRICANE Last Direct Hit By County	State	City (County)	MAJOR HURRICANE Return Period	MAJOR HURRICANE Last Direct Hit By County	HURRICANE Return Period	HURRICANE Last Direct Hit By County
Texas	Brownsville (Cameron)	25 yrs	1980(3) Allen	11 yrs	1980(3) Allen	Florida	Vero Beach (Indian River)	18 yrs	2004(3) Jeanne	7 yrs	2004(3) Jeanne
	Corpus Christi (Nueces)	24	1970(3) Celia	12	1971(1) Fern		Cocoa Beach (Brevard)	22	2004(3) Jeanne	8	2004(3) Jeanne
	Port Aransas (Aransas)	23	1970(3) Celia	11	1971(1) Fern		Daytona Beach (Volusia)	31	<1880	8	1960(2) Donna
	Matagorda (Matagorda)	19	1961(4) Carla	9	2003(1) Claudette		St. Augustine (St. Johns)	29	<1880	8	1964(2) Dora
	Freeport (Brazoria)	19	1983(3) Alicia	7	1983(3) Alicia		Jacksonville (Duval)	28	<1880	9	1964(2) Dora
	Galveston (Galveston)	18	1983(3) Alicia	11	1989(1) Jerry		Fernandina Beach (Nassau)	33	<1880	8	1928(2)
	Houston (Harris)	21	1941(3)	11	1989(1) Jerry	Georgia	Brunswick (Camden)	31	1898(4)	8	1928(1)
	Beaumont (Jefferson)	25	2005(3) Rita	12	2005(3) Rita		Savannah (Chatham)	34	1854(3)	8	1979(2) David
Louisiana	Cameron (Cameron)	24	2005(3) Rita	10	2005(3) Rita	S. Carolina	Hilton Head (Beaufort)	24	1959(3) Gracie	7	1979(2) David
	Morgan City (St. Mary)	20	1992(3) Andrew	8	2002(1) Lili		Charleston (Charleston)	15	1989(4) Hugo	6	2004(1) Gaston
	Houma (Terrebonne)	18	1992(3) Andrew	7	1992(3) Andrew		Myrtle Beach (Horry)	16	1954(4) Hazel	6	2004(1) Charley
	New Orleans (Orleans)	19	2005(3) Katrina	8	2005(3) Katrina	N. Carolina	Wilmington (New Hanover)	16	1996(3) Fran	5	2005(1) Ophelia
Mississippi	Bay St. Louis (Hancock)	24	2005(3) Katrina	10	2005(3) Katrina		Morehead City (Carteret)	13	1996(3) Fran	4	2005(1) Ophelia
	Biloxi (Harrison)	18	2005(3) Katrina	8	2005(3) Katrina		Cape Hatteras (Dare)	11	1993(3) Emily	3	2003(2) Isabel
	Pascagoula (Jackson)	19	2005(3) Katrina	7	2005(3) Katrina	Virginia	Virginia Beach (Virginia Beach)	36	1944(3)	7	2003(1) Isabel
Alabama	Mobile (Mobile)	23	2004(3) Ivan	10	2004(3) Ivan		Norfolk (Norfolk)	43	<1851	10	2003(1) Isabel
	Gulf Shores (Baldwin)	17	2004(3) Ivan	6	2004(3) Ivan	Maryland	Ocean City (Worcester)	48	<1851	12	1878(1)
	Pensacola (Escambia)	17	2005(3) Dennis	7	2005(3) Dennis		Baltimore (Baltimore)	>500	<1851	56	1878(1)
Florida	Destin (Okaloosa)	19	1995(3) Opal	6	1995(3) Opal	Delaware	Rehoboth Beach (Sussex)	55	<1851	13	1903(1)
	Panama City (Bay)	17	1995(3) Opal	6	1995(3) Opal		Wilmington (New Castle)	>500	<1851	37	1878(1)
	Apalachicola (Franklin)	33	1985(3) Elena	7	1998(2) Earl	New Jersey	Cape May (Cape May)	64	<1851	14	1903(1)
	Homosassa (Citrus)	26	1950(3) Easy	8	1968(2) Gladys		Atlantic City (Atlantic)	69	<1851	13	1903(1)
	St. Petersburg (Pinellas)	19	1921(3)	6	1946(1)	New York	New York City (New York)	150	<1851	18	1903(1)
	Tampa (Hillsboro)	23	1921(3)	6	1946(1)		Westhampton (Suffolk)	62	1985(3) Gloria	13	1985(3) Gloria
	Sarasota (Sarasota)	19	1944(3)	6	1946(1)	Connecticut	New London (New London)	54	1938(3)	14	1991(2) Bob
	Fort Myers (Lee)	15	2004(4) Charley	6	2004(4) Charley		New Haven (New Haven)	86	1938(3)	18	1985(2) Gloria
	Naples (Collier)	14	2005(3) Wilma	6	2005(3) Wilma		Bridgeport (Fairfield)	84	1954(3) Carol	18	1985(2) Gloria
	Key West (Monroe)	12	2005(3) Wilma	5	2005(3) Wilma	Rhode Island	Providence (Providence)	77	1954(3) Carol	16	1991(2) Bob
	Miami (Miami-Dade)	9	1992(5) Andrew	4	2005(2) Wilma	Mass.	Cape Cod (Barnstable)	42	1954(3) Edna	10	1991(2) Bob
	Fort Lauderdale (Broward)	10	1950(3) King	4	2005(2) Wilma		Boston (Suffolk)	170	1869(3)	22	1960(1) Donna
	W. Palm Beach (Palm Beach)	13	2004(3) Jeanne	6	2005(2) Wilma	N. Hampshire	Portsmouth (Rockingham)	>500	<1851	28	1985(2) Gloria
	Stuart (Martin)	15	2004(3) Jeanne	6	2005(2) Wilma	Maine	Portland (Cumberland)	>500	<1851	33	1985(1) Gloria
	Fort Pierce (St. Lucie)	17	2004(3) Jeanne	7	2004(3) Jeanne		Eastport (Washington)	160	<1851	19	1969(1) Gerda

Notes: <1900 means before 1900 etc.

Table 13a. Estimated annual deaths and damages (unadjusted and adjusted for inflation[1] and normalized[L] for inflation, growth in personal wealth and population) in the mainland United States from landfalling Atlantic or Gulf tropical cyclones 1900-2006.

Year	Deaths	Unadjusted	Adjusted[1]	Normalized[L]	Year	Deaths	Unadjusted	Adjusted[1]	Normalized[L]
1900	8,000 [+]	30	1,358 [2]	104,330	1954	193	756	7,126	37,455
1901	10	1	45 [2]	213	1955	218	985	9,119	24,438
1902	0	Minor	Minor	-	1956	19	27	236	606
1903	15	1	45 [2]	6,803	1957	426	152	1,292	4,034
1904	5	2	91 [2]	1,139	1958	2	11	94	535
1905	0	Minor	Minor	-	1959	24	23	198	902
1906	298	3 [+]	136 [2]	4,080	1960	65	396	3,423	31,469
1907	0	Minor	Minor	-	1961	46	414	3,593	15,192
1908	0	Minor	Minor	-	1962	3	2	17	97
1909	406	8	362 [2]	3,081	1963	10	12	101	259
1910	30	1	45 [2]	876	1964	49	515	4,452	16,478
1911	17	1 [+]	45 [2]	235	1965	75	1,445	12,088	22,324
1912	1	Minor	Minor	-	1966	54	15	120	353
1913	5	3	136 [2]	724	1967	18	200	1,541	4,217
1914	0	Minor	Minor	-	1968	9	10	73	690
1915	550	63	2,853 [3]	74,262	1969	256	1,421	9,784	22,286
1916	107	33	1,227	7,919	1970	11	454	3,045	5,909
1917	5	Minor	Minor	-	1971	8	213	1,353	2,188
1918	34	5	121	886	1972	122	2,100	12,424	18,458
1919	287 [4]	22	477	14,392	1973	5	18	97	153
1920	2	3	51	367	1974	1	150	736	1,127
1921	6	3	64	3,348	1975	21	490	2,210	2,931
1922	0	Minor	Minor	-	1976	9	100	424	511
1923	0	Minor	Minor	-	1977	0	10	38	56
1924	2	Minor	Minor	-	1978	36	20	68	153
1925	6	Minor	Minor	-	1979	22	3,045	9,164	14,801
1926	408	112	2,405	169,398	1980	2	300	819	1,682
1927	0	Minor	Minor	-	1981	0	25	64	180
1928	2,500	25	537	35,298	1982	0	Minor	Minor	45
1929	3	1	20	390	1983	22	2,000	4,825	7,843
1930	0	Minor	Minor	-	1984	4	66	153	304
1931	0	Minor	Minor	-	1985	30	4,000	9,113	11,622
1932	40	8	183	6,210	1986	9	17	37	53
1933	63	47	1,194	14,006	1987	0	8	17	20
1934	17	5	116	932	1988	6	59	118	182
1935	414	12	278	9,150	1989	56	7,670	14,770	17,609
1936	9	2	48	838	1990	13	57	106	133
1937	0	Minor	Minor	-	1991	16	1,500	2,775	3,196
1938	600	306	6,571	41,140	1992	24	26,500	48,058	60,547
1939	3	Minor	Minor	-	1993	4	57	99	133
1940	51	5	112	1,224	1994	38	973	1,611	2,036
1941	10	8	167	2,530	1995	29	3,723	5,905	7,877
1942	8	27	489	2,475	1996	36	3,600	5,602	6,864
1943	16	17	288	3,746	1997	4	100	151	172
1944	64	165	2,794	54,760	1998	23	4,344	6,401	6,323
1945	7	80	1,323	14,676	1999	62	5,532	7,797	8,692
1946	0	5	70	4,953	2000	6	27	36	38
1947	53	136	1,600	20,071	2001	45	5,260	6,747	7,319
1948	3	18	192	4,249	2002	9	1,220	1,522	1,566
1949	4	59	630	16,147	2003	24	3,600	4,257	4,423
1950	19	36	379	5,806	2004	60	45,000	48,965	51,587
1951	0	2	18	376	2005	1525	115,520	120,718	121,296
1952	3	3	28	120	2006	0	500	500	500
1953	2	6	57	59					

[+] 1900 could have been as high as 12,000, other years means "more than".
[1] Adjusted to 2006 dollars based on U.S. Department of Commerce Implicit Price Deflator for Construction.
[2] Using 1915 cost adjustment - none available prior to 1915.
[3] Considered too high in 1915 reference.
[4] Could include some offshore losses.
[L] Normalization reflects inflation, changes in personal wealth and coastal county population to 2005, (Pielke et al. 2007.) then including an estimate to 2006 dollars by increasing totals by 5%.

Table 13b. As in Table 13a, but for the thirty deadliest years from 1851-2006 and costliest years from 1900 to 2006.

	Ranked on Deaths		Ranked on Unadjusted Damage		Ranked on Adjusted[1] Damage		Ranked by Normalized[L] Damage	
	Year	Deaths	Year	($ Millions)	Year	($ Millions)	Year	($ Millions)
1	1900	8,000 [+]	2005	115,520	2005	120,718	1926	169,398
2	1893 ~	3,000 [s]	2004	45,000	2004	48,965	2005	121,296
3	1928	2,500	1992	26,500	1992	48,058	1900	104,330
4	2005	1,525	1989	7,670	1989	14,770	1915	74,262
5	1881	700	1999	5,532	1972	12,424	1992	60,547
6	1915	550	2001	5,260	1965	12,088	1944	54,760
7	1957	426	1998	4,344	1969	9,784	2004	51,587
8	1935	414	1985	4,000	1979	9,164	1938	41,140
9	1926	408	1995	3,723	1955	9,119	1954	37,455
10	1909	406	1996	3,600	1985	9,113	1928	35,298
11	1906	298	2003	3,600	1999	7,797	1960	31,469
12	1919	287 [s]	1979	3,045	1954	7,126	1955	24,438
13	1969	256	1972	2,100	2001	6,747	1965	22,324
14	1938	256	1983	2,000	1938	6,571	1969	22,286
15	1955	218	1991	1,500	1998	6,401	1947	20,071
16	1954	193	1965	1,445	1995	5,905	1972	18,458
17	1972	122	1969	1,421	1996	5,602	1989	17,609
18	1916	107	2002	1,220	1983	4,825	1964	16,478
19	1965	75	1955	985	1964	4,452	1949	16,147
20	1960	65	1994	973	2003	4,257	1961	15,192
21	1944	64	1954	756	1961	3,593	1979	14,801
22	1933	63	1964	515	1960	3,423	1945	14,676
23	1999	62	2006	500	1970	3,045	1919	14,392
24	2004	60	1975	490	1915	2,853 [2]	1933	14,006
25	1989	56	1970	454	1944	2,794	1985	11,622
26	1966	54	1961	414	1991	2,775	1935	9,150
27	1947	53	1960	396	1926	2,405	1999	8,692
28	1940	51	1938	306	1975	2,210	1916	7,919
29	1964	49	1980	300	1994	1,611	1995	7,877
30	1961	46	1971	213	1947	1,600	1983	7,843

[+] Could have been as high as 12,000.
[1] Adjusted to 2006 dollars based on U.S. Department of Commerce Implicit Price Deflator for Construction.
[2] Considered too high in 1915 reference.
[3] Using 1915 cost adjustment - none available prior to 1915.
[s] Could include offshore losses
[L] Normalization reflects inflation, changes in personal wealth and coastal county population to 2005, (Pielke et al. 2007) then including an estimate to 2006 dollars by increasing totals by 5%.

Table 14. Deadliest and Costliest Hurricanes from 1900 to 2006 to affect Hawaii, Puerto Rico and the U.S. Virgin Islands.

Name	Date	Island or CPA	Unadjusted Damage ($000)	Adjusted for Inflation [3]	Deaths	Max Wind (Mph)	Min P (Mb)
Mokapu Cyclone	Aug 19, 1938	25 mi NE Oahu	Unk	Unk	Unk	Unk	Unk
Hiki	Aug 15, 1950	100 mi NE Hawaii	Unk	Unk	Unk	Unk	Unk
Nina	Dec 02, 1957	100 mi SW Kauai	200	1,700	4	90	965
Dot	Aug 06, 1959	Kauai	6,000	51,660	0	115	955
Iwa	Nov 23, 1982	25 mi NW Kauai	312,000	773,760	1	90	964
Iniki	Sep 11, 1992	Kauai	1,800,000	3,258,000	4	130	950
San Hipolito	Aug 22, 1916	Puerto Rico	1,000	37,196	1	100	988
San Liborio	Jul 23, 1926 [1]	SW Puerto Rico	5,000	107,371	25	60	~985
San Felipe	Sep 13, 1928	Puerto Rico	85,000	1,825,308	312	160	Unk
San Nicolas	Sep 10, 1931 [1]	Puerto Rico	200	4,578	2	120	Unk
San Ciprian	Sep 26, 1932 [1]	USVI, PR	30,000	686,703	225	100	948
San Mateo	Sep 21, 1949	St. Croix	Unk	-	Unk	80	~985
Santa Clara (Betsy)	Aug 12, 1956	Puerto Rico	40,000	350,084	16	90	991
Donna	Sep 05, 1960 [1]	PR & St. Thomas	Unk	-	107	135	958
Eloise (T.S.)	Sep 15, 1975 [1]	Puerto Rico	Unk	-	44	40	1007
David	Aug 30, 1979 [2]	S. of Puerto Rico	Unk	-	Unk	175	924
Frederic (T.S.)	Sep 04, 1979 [2]	Puerto Rico	125,000	376,209	7	60	1000
Hugo	Sep 18, 1989	USVI, PR	1,000,000	1,925,743	5	140	940
Marilyn	Sep 16, 1995	USVI, E. PR	1,500,000	2,379,205	8	110	952
Hortense	Sep 10, 1996	SW Puerto Rico	500,000	778,000	18	80	989
Georges	Sep 21, 1998	USVI & PR	1,800,000	2,652,273	0	115	968
Lenny	Nov 17, 1999	USVI & PR	330,000	465,109	0	155	933

[1] Effects continued into the following day. [2] Damage and Casualties from David and Frederic are combined.
[3] Adjusted to 2006 dollars based on U.S. Department of Commerce Implicit Price Deflator for Construction

(14) Are there hurricane landfall cycles? Figures 1 through 16 show the landfalling portion of the tracks of major hurricanes that have struck the United States between 1851-2006. The reader might note the tendency for the major hurricane landfalls to cluster in certain areas during certain decades. Another interesting point is the tendency for this clustering to occur in the latter half of individual decades in one area and in the first half of individual decades in another area. During the very active period of the thirties this clustering is not apparent.

A comparison of twenty-year periods beginning in 1851 indicates that the major hurricanes tended to be in Gulf Coast states before 1891, then favored Florida and the western Gulf until 1911, shifting to the eastern Gulf Coast states and Florida during the next twenty years, then to Florida and the Atlantic Coast states during the 1940s-1950s, and back to the western Gulf Coast states in the following twenty-year period. Most major hurricanes have recently favored Florida and the central Gulf Coast states.

CONCLUSIONS

In virtually every coastal city from Texas to Maine, the present National Hurricane Center Director (Bill Proenza) and former directors have stated that the United States is building toward its next hurricane disaster. Hurricane Katrina is a sad reminder of the vulnerability of the United States to hurricanes. The areas along the United States Gulf and Atlantic coasts where most of this country's hurricane related fatalities have occurred are also experiencing the country's most significant growth in population. Low hurricane experience levels, as shown by Hebert et al. (1984), Jarrell et al. (1992) and Table 12, are a serious problem and could lead to future disasters. This situation, in combination with continued building along the coast, will lead to dangerous problems for many areas in hurricanes. Because it is likely that people will always be attracted to live along the shoreline, a solution to the problem lies in education and preparedness as well as long-term policy and planning.

The message to coastal residents is this: Become familiar with what hurricanes can do, and when a hurricane threatens your area, increase your chances of survival by moving away from the water until the hurricane has passed! Unless this message is clearly understood by coastal residents through a thorough and continuing preparedness effort, disastrous loss of life is inevitable in the future.

Acknowledgments: Richard Pasch and Colin McAdie made helpful suggestions and Michelle Mainelli assisted with producing some of the tables. Paul Hebert, Glenn Taylor, Bob Case, Max Mayfield and Jerry Jarrell, co-authors of previous versions of this paper, are recognized for their enduring contributions to this work. David Roth provided the source for the Audrey update, and Joan David drafted the decade-by-decade major hurricane figures.

REFERENCES

Blake, E.S., E.N. Rappaport, J.D. Jarrell, and C.W. Landsea, 2005: The Deadliest, Costliest and Most Intense United States Hurricanes from 1851 to 2004 (and Other Frequently Requested Hurricane Facts). NOAA, Technical Memorandum NWS-TPC-4, 48 pp.

Gentry, R.C., 1966: Nature and Scope of Hurricane Damage. American Society for Oceanography, Hurricane Symposium, Publication Number One, 344 pp.

Hebert, P.J. and J.G. Taylor, 1975: Hurricane Experience Levels of Coastal County Populations - Texas to Maine. *Special Report*, National Weather Serivce Community Preparedness Staff and Southern Region, July, 153 pp.

Hebert, P.J., J.G. Taylor, and R.A. Case, 1984: Hurricane Experience Levels of Coastal County Populations - Texas to Maine. NOAA, Technical Memorandum NWS-NHC-24, 127 pp.

Hebert, P.J., J.D. Jarrell, and B.M. Mayfield, 1997: The Deadliest, Costliest and Most Intense United States Hurricanes of This Century (and Other Frequently Requested Hurricane Facts). NOAA, Technical Memorandum NWS-TPC-1, 30 pp.

Jarrell, J.D., B.M. Mayfield, E.N. Rappaport, and C.W. Landsea, 2001: The Deadliest, Costliest and Most Intense United States Hurricanes from 1900 to 2000 (and Other Frequently Requested Hurricane Facts). NOAA, Technical Memorandum NWS-TPC-3, 30 pp.

Jarrell, J.D., P.J. Hebert, and B.M. Mayfield, 1992: Hurricane Experience Levels of Coastal County Populations - Texas to Maine. NOAA, Technical Memorandum NWS-NHC-46, 152 pp.

Jarvinen, B.R., C.J. Neumann, and A.S. Davis, 1984: A Tropical Cyclone Data Tape for the North Atlantic Basin, 1886-1983: Contents, Limitations, and Uses. NOAA, Technical Memorandum NWS-NHC-22, 21 pp.

Landsea, C.W. et al, 2004b: The Atlantic Hurricane Database Reanalysis Project. Documentation for 1851-1910 alterations and additions to the HURDAT database. *Hurricanes and Typhoons: Past, Present and Future*, R.J. Murnane and K.B. Liu, Eds., Columbia University Press, 177-221.

Neumann, C. J., 1987: The National Hurricane Center Risk Analysis Program (HURISK). NOAA Technical Memorandum, NWS NHC 38, 56 pp.

Neumann, C.J., B.R. Jarvinen, C.J. McAdie, and G.R. Hammer, 1999: Tropical Cyclones of the North Atlantic Ocean, 1871-1998. NOAA, Historical Climatology Series 6-2, 206 pp.

Pielke, Jr., R.A., and C.W. Landsea, 1998: Normalized U.S. Hurricane Damage. 1925-1995, *Weather & Forecasting*, 13, 621-631.

Pielke, Jr., R.A., J. Gratz, C.W. Landsea, D. Collins, M. Saunders, and R. Musulin, 2007: Normalized Hurricane Damages in the United States: 1900-2005. *Natural Hazards Review*, (Submitted).

Ross, N.M.W. and S.M. Goodson: Hurricane Audrey. Sulphur Louisiana, Wise Publications; 1997.

Simpson, R.H., 1974: The hurricane disaster potential scale. *Weatherwise*, Vol. 27, 169-186.

U.S. Weather Bureau: Climatological Data and Storm Data, various volumes, various periods, National and State Summaries (National Weather Service 1971-1998).

U.S. Weather Bureau: *Monthly Weather Review*, 1872-1970 (National Weather Service 1971-1973, and American Meteorological Society 1974-2004).

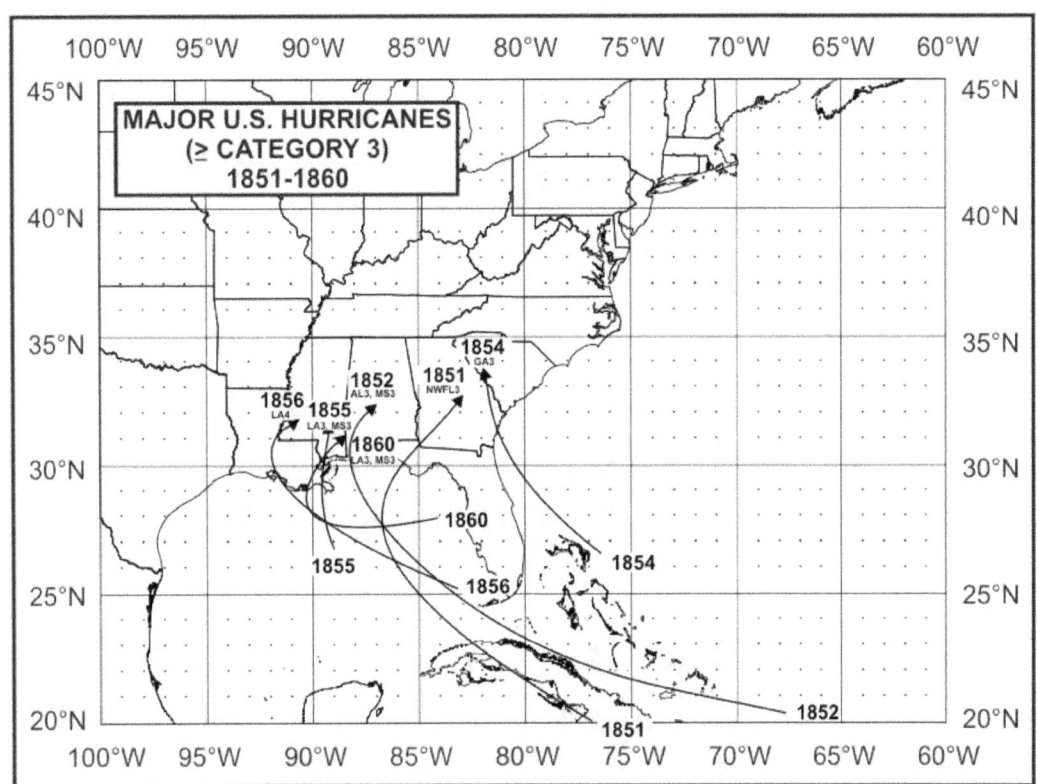

Figure 1. United States major hurricane strikes (stronger than or equal to a category 3) during the period 1851-1860.

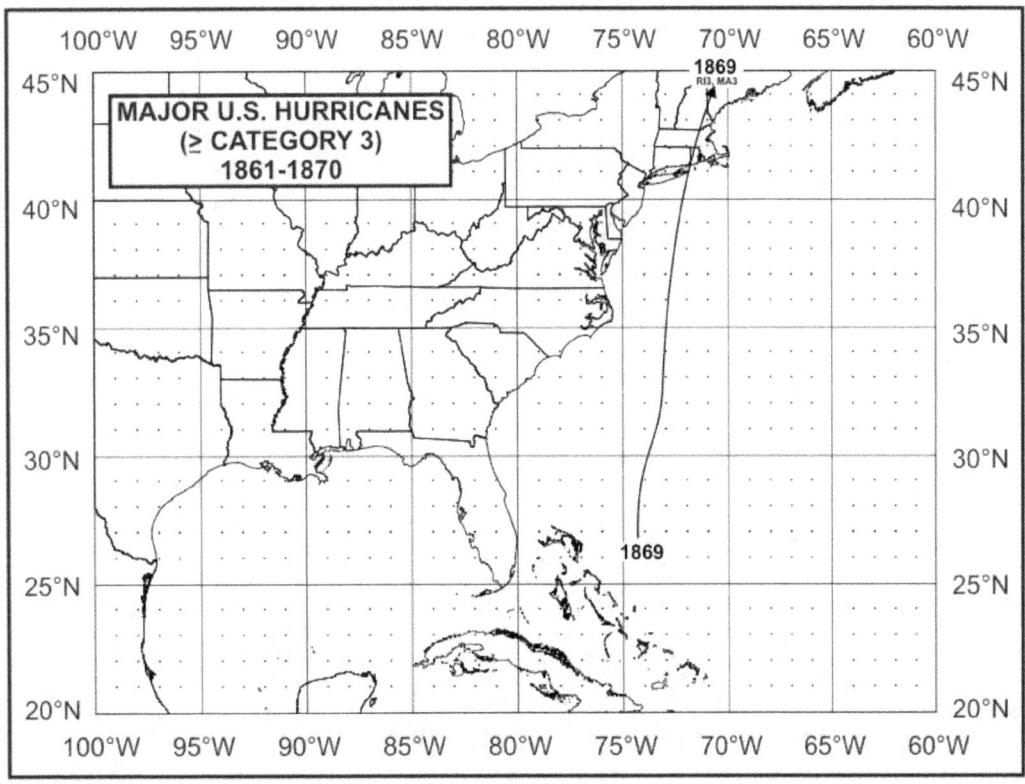

Figure 2. United States major hurricane strikes (stronger than or equal to a category 3) during the period 1861-1870.

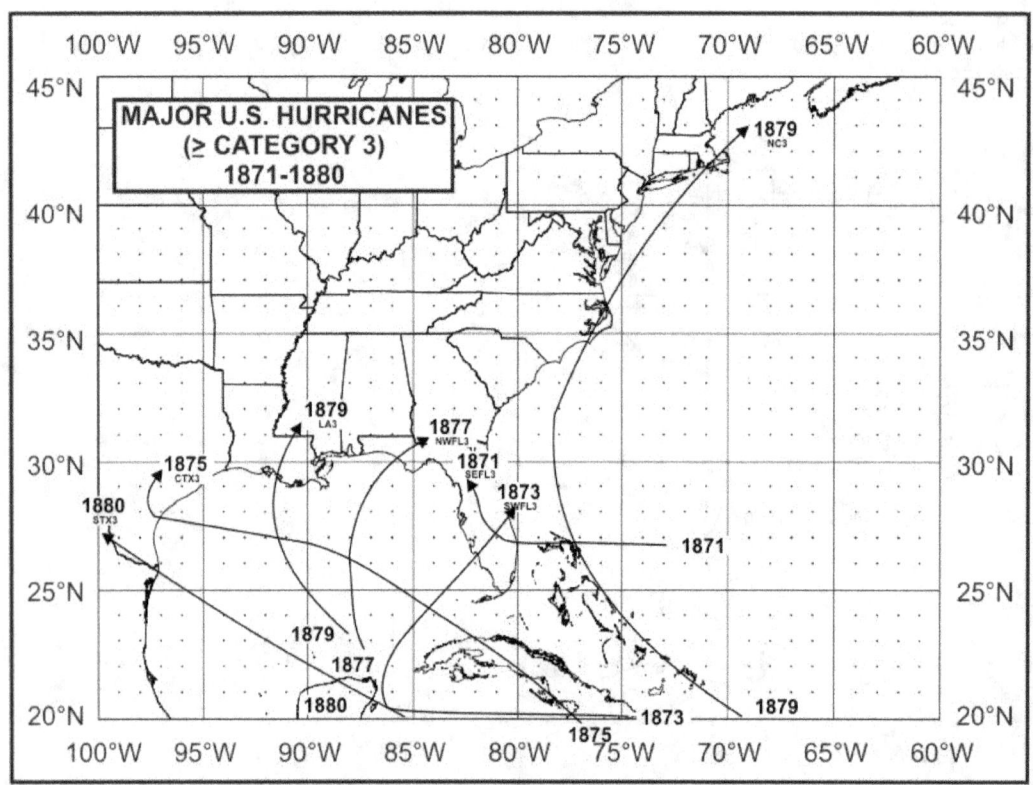

Figure 3. United States major hurricane strikes (stronger than or equal to a category 3) during the period 1871-1880.

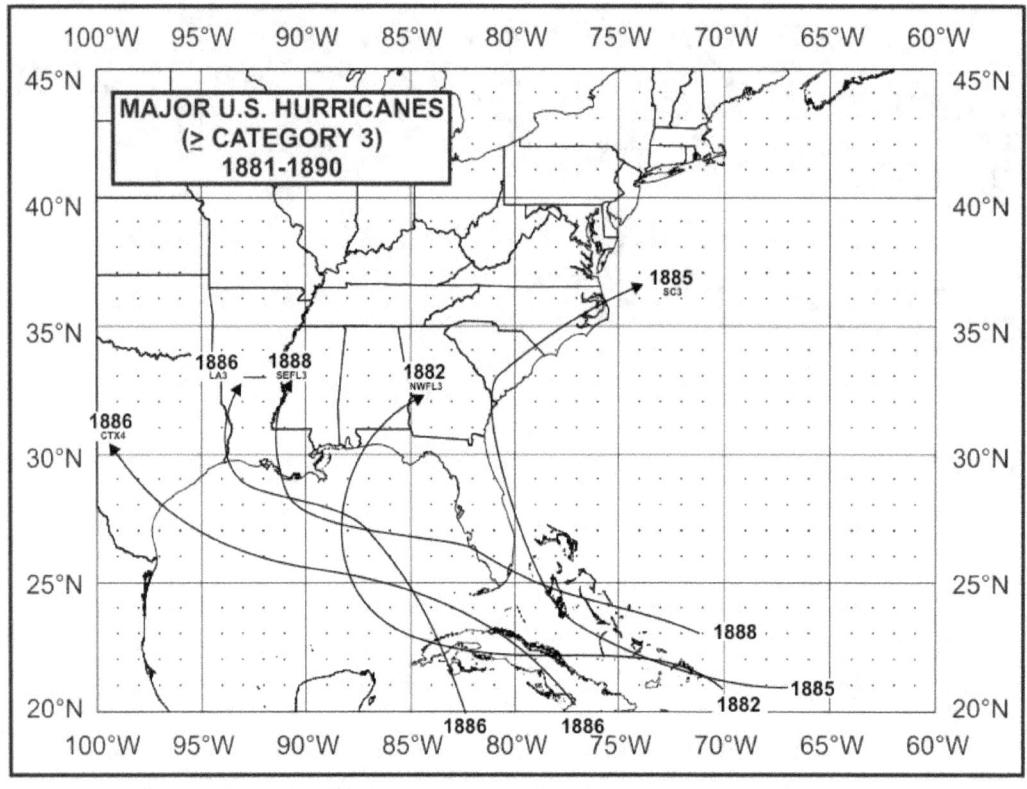

Figure 4. United States major hurricane strikes (stronger than or equal to a category 3) during the period 1881-1890.

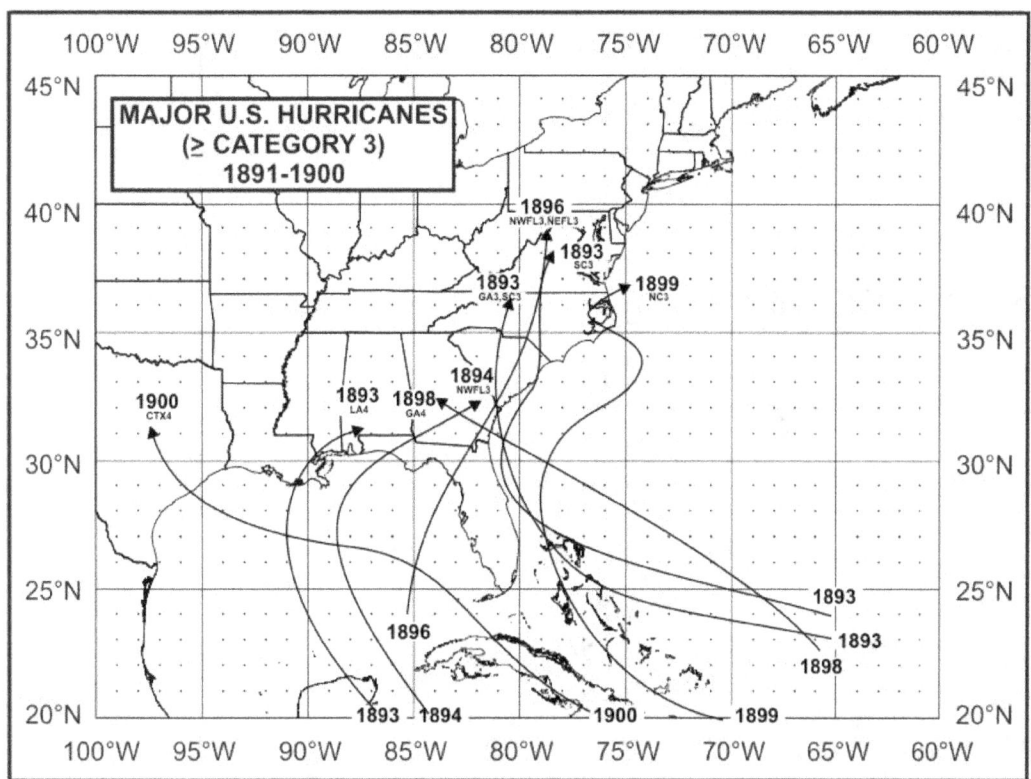

Figure 5. United States major hurricane strikes (stronger than or equal to a category 3) during the period 1891-1900.

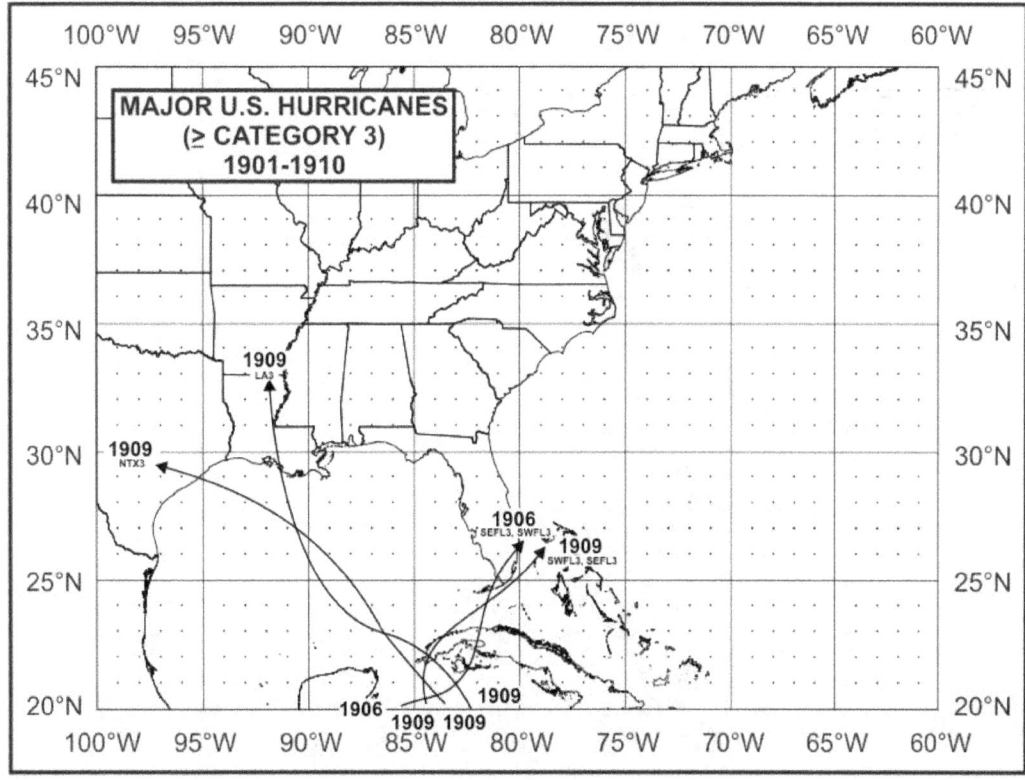

Figure 6. United States major hurricane strikes (stronger than or equal to a category 3) during the period 1901-1910.

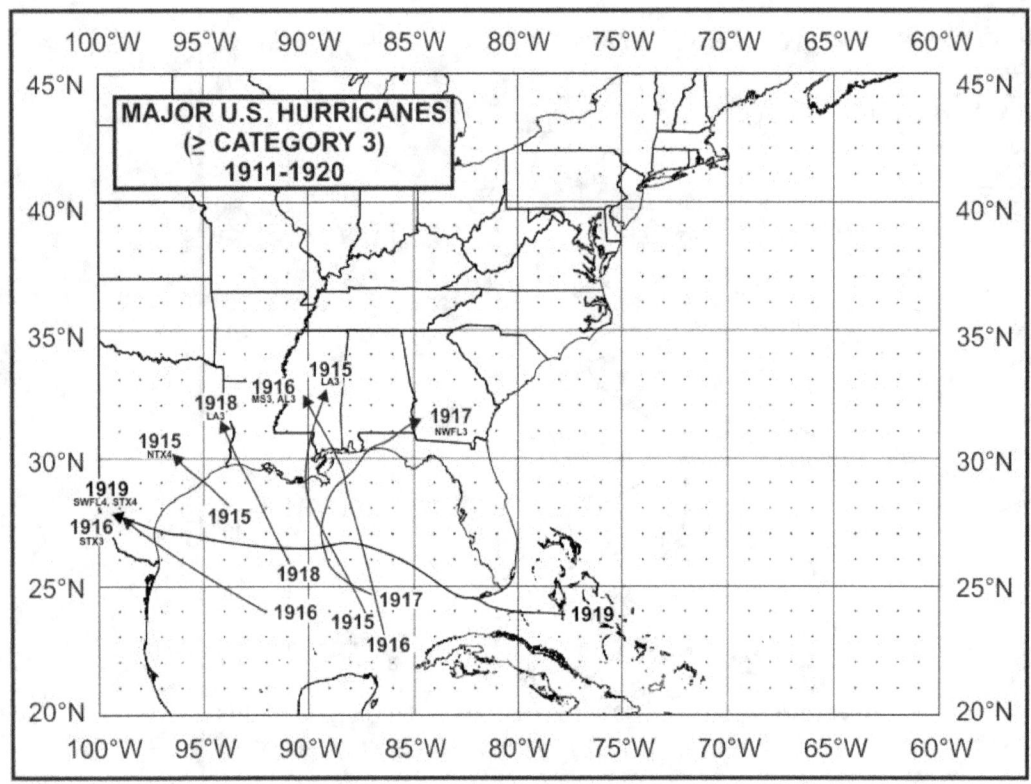

Figure 7. United States major hurricane strikes (stronger than or equal to a category 3) during the period 1911-1920.

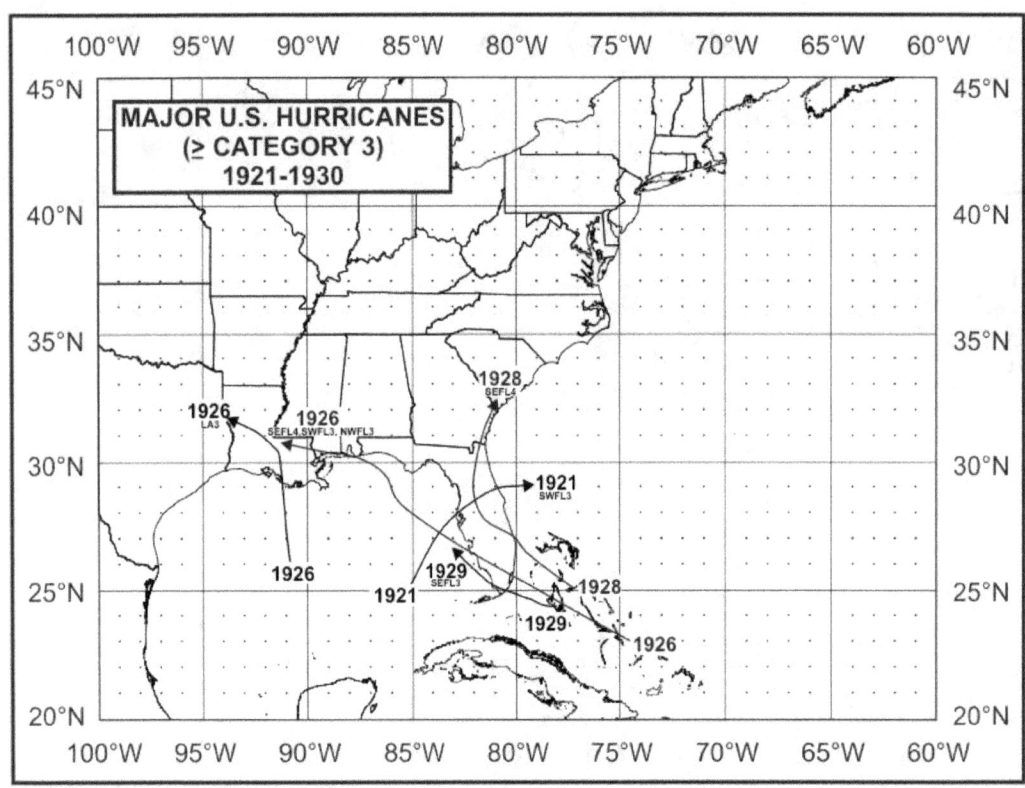

Figure 8. United States major hurricane strikes (stronger than or equal to a category 3) during the period 1921-1930.

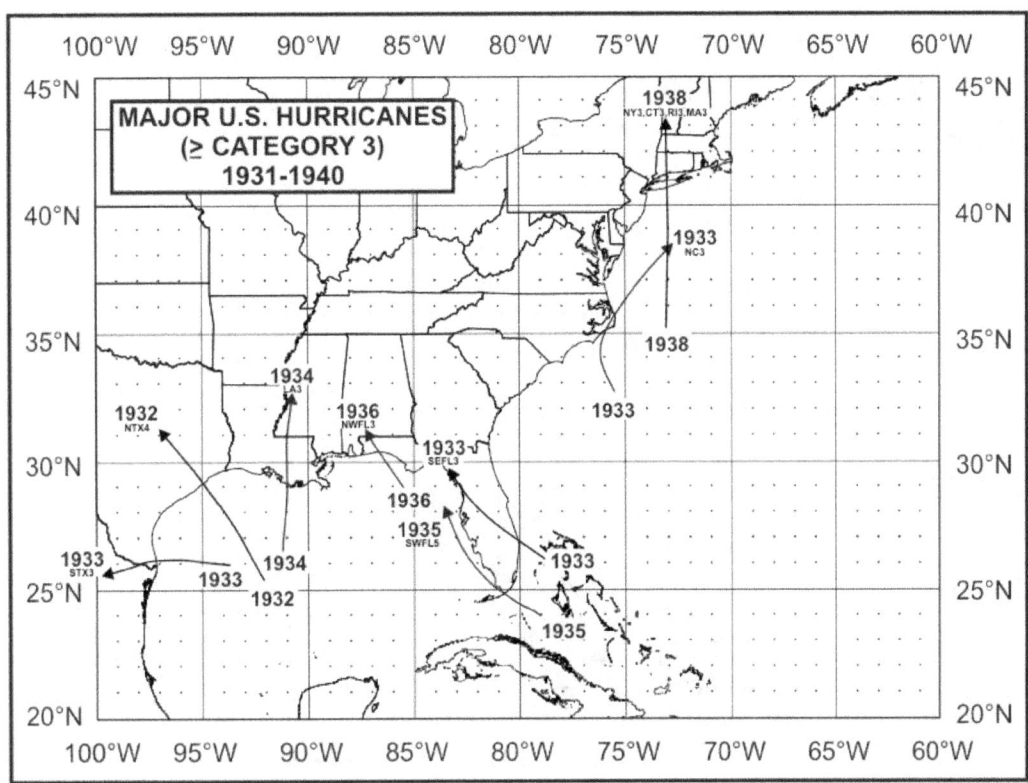

Figure 9. United States major hurricane strikes (stronger than or equal to a category 3) during the period 1931-1940.

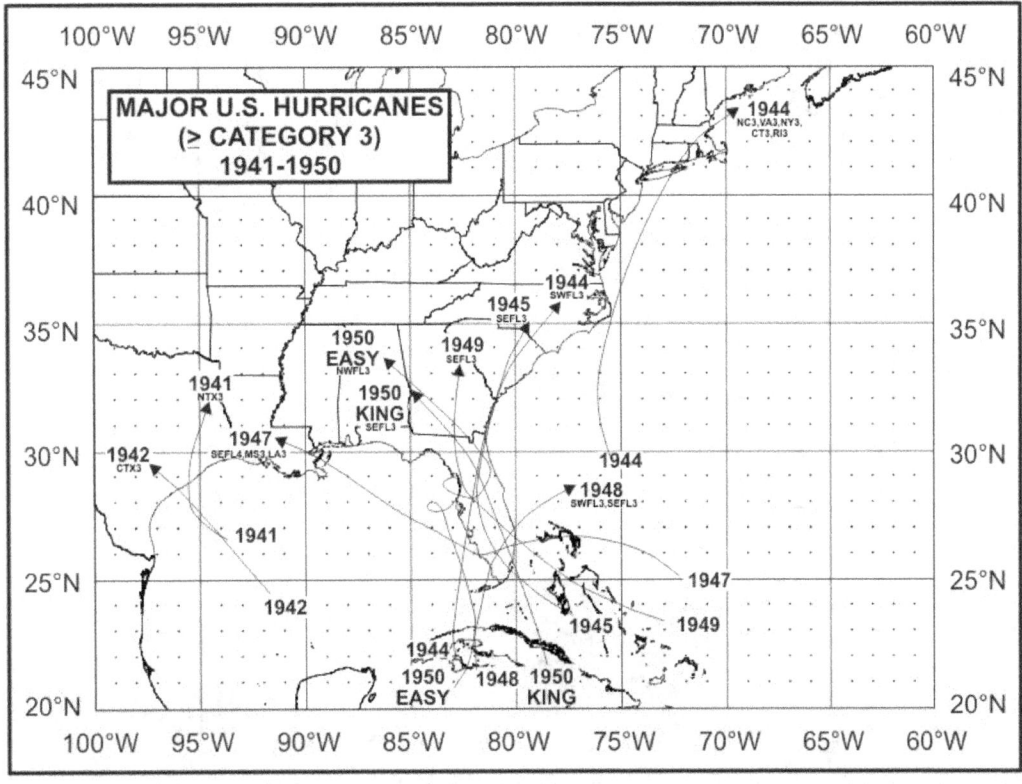

Figure 10. United States major hurricane strikes (stronger than or equal to a category 3) during the period 1941-1950.

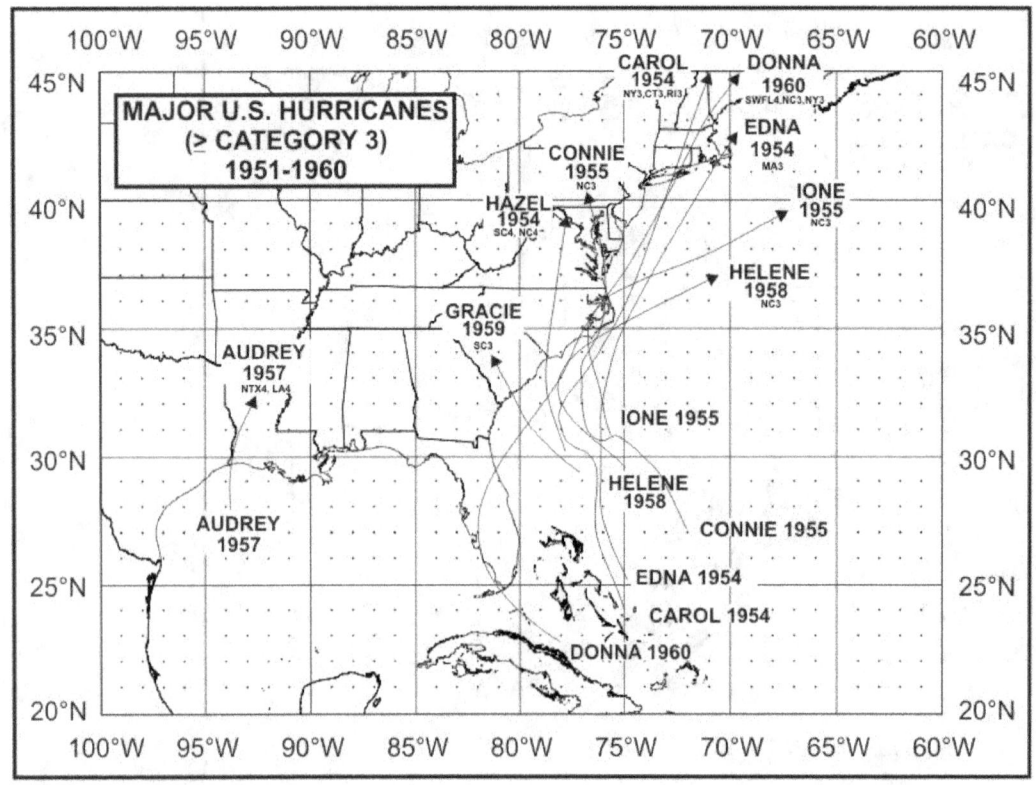

Figure 11. United States major hurricane strikes (stronger than or equal to a category 3) during the period 1951-1960.

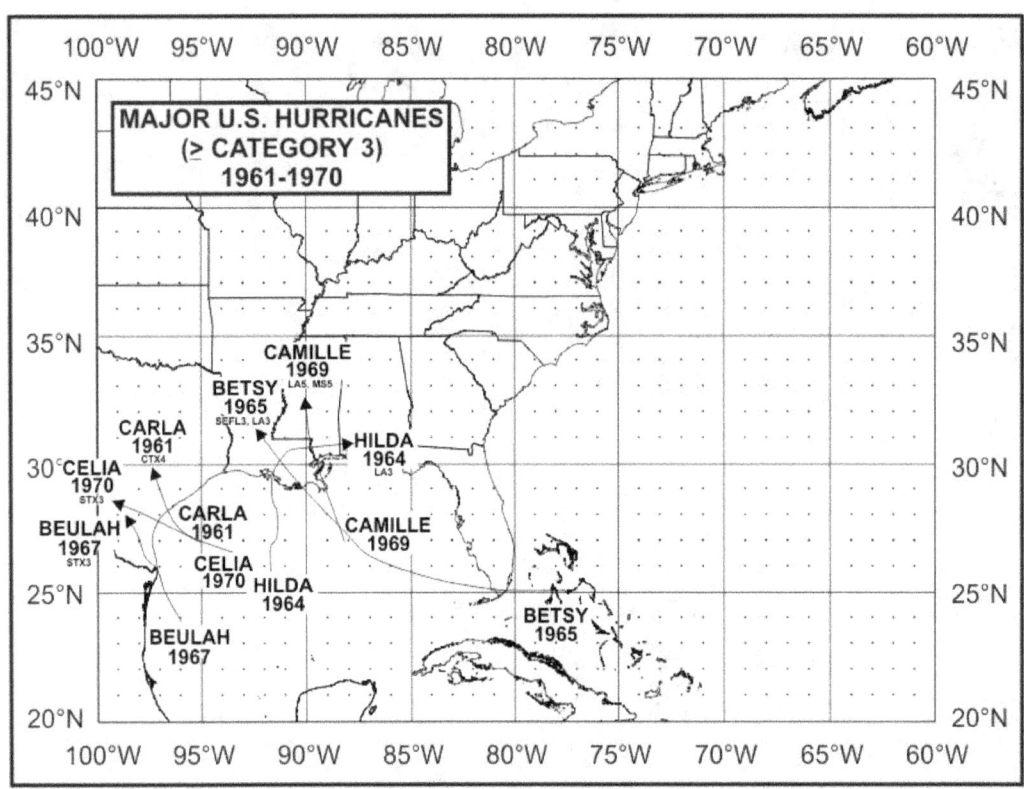

Figure 12. United States major hurricane strikes (stronger than or equal to a category 3) during the period 1961-1970.

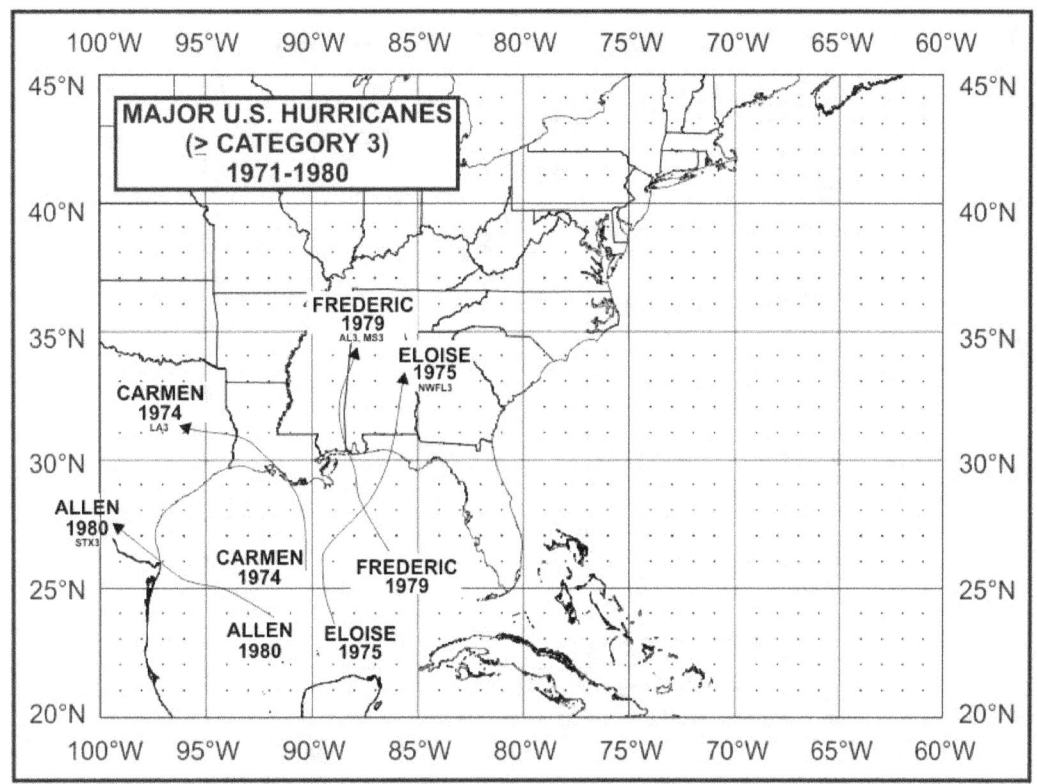

Figure 13. United States major hurricane strikes (stronger than or equal to a category 3) during the period 1971-1980.

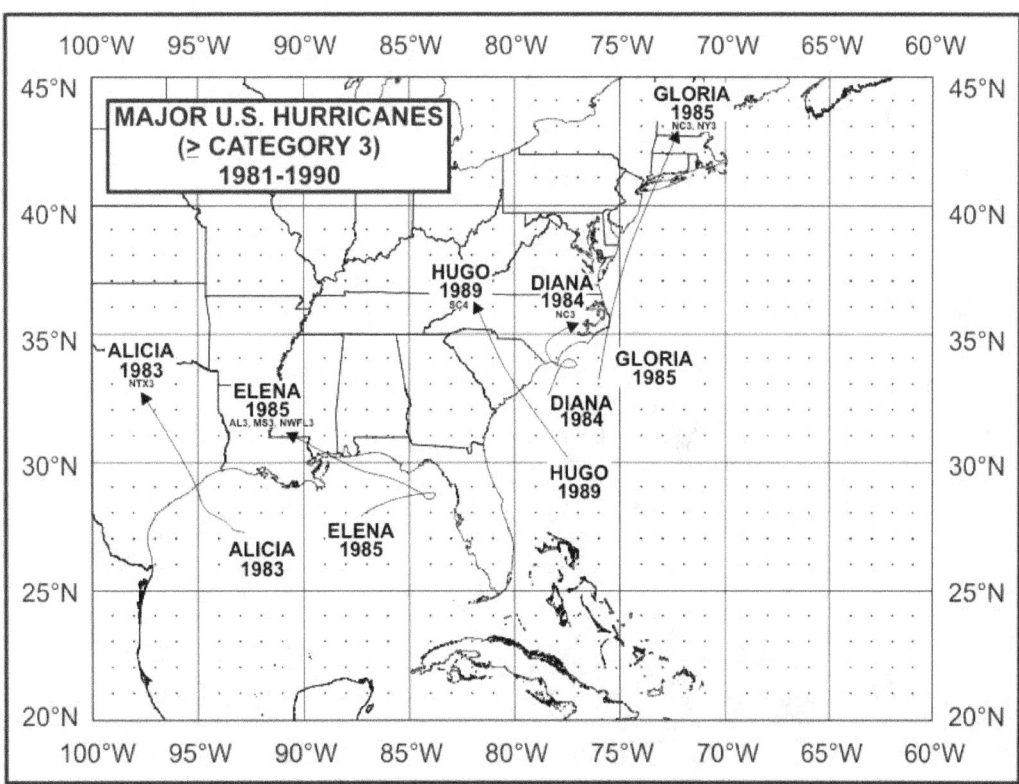

Figure 14. United States major hurricane strikes (stronger than or equal to a category 3) during the period 1981-1990.

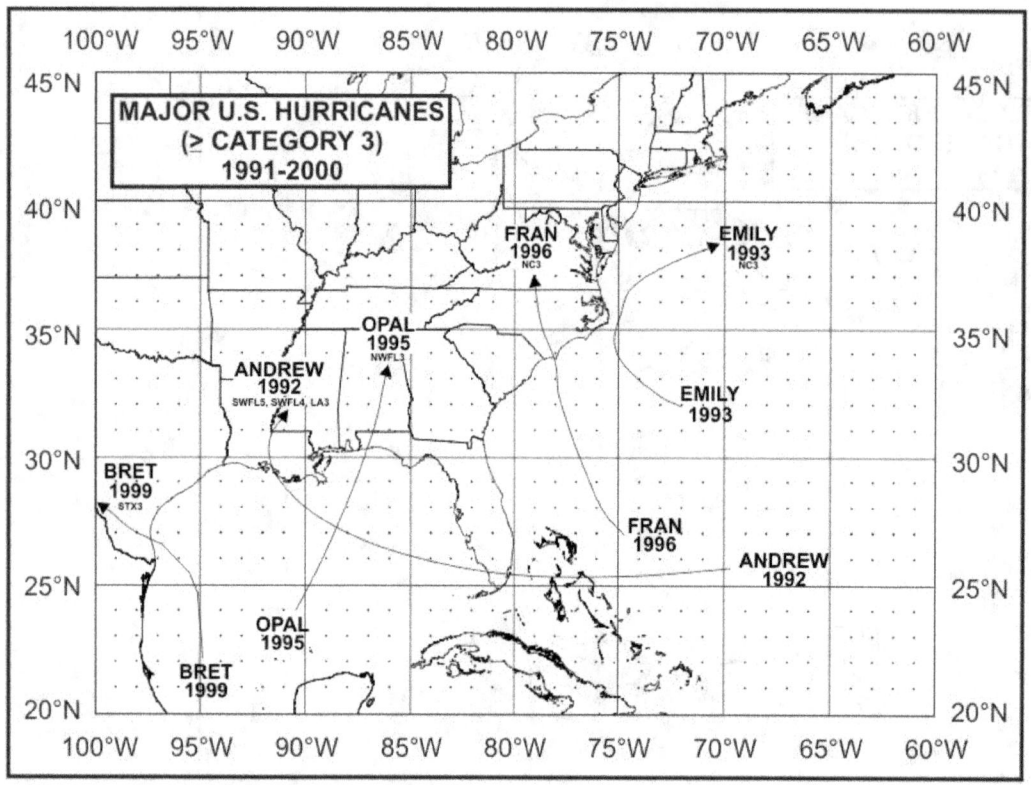

Figure 15. United States major hurricane strikes (stronger than or equal to a category 3) during the period 1991-2000.

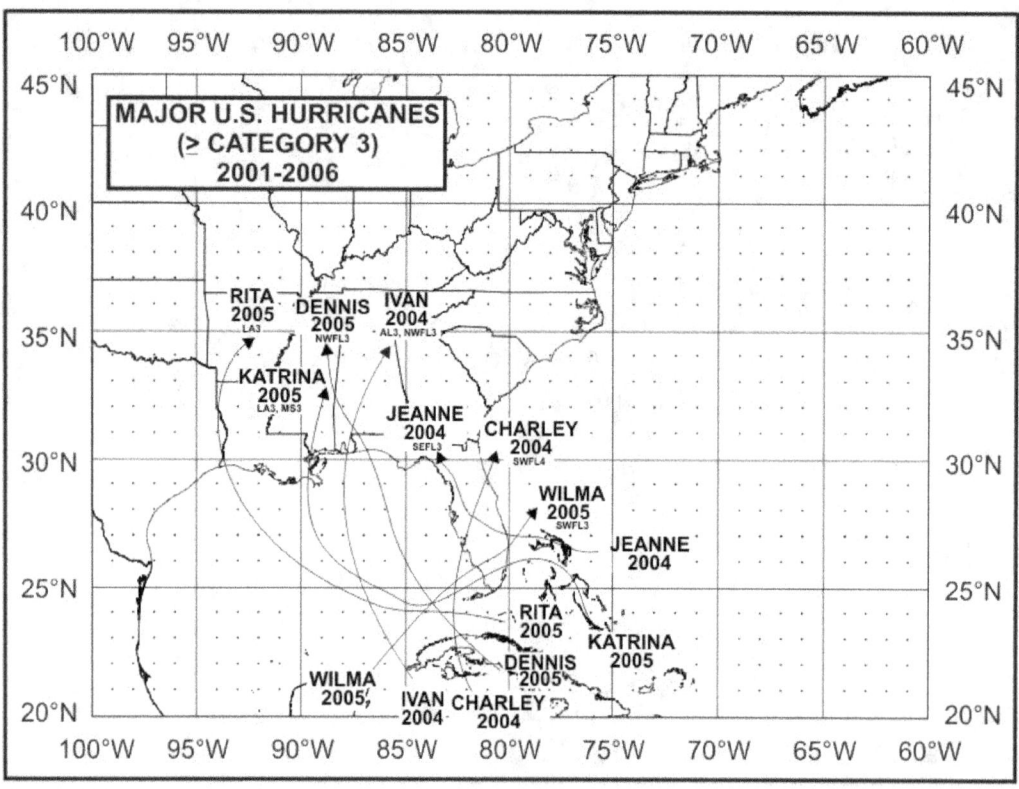

Figure 16. United States major hurricane strikes (stronger than or equal to a category 3) during the period 2001-2006.

Appendix A: Chronological List of All Hurricanes which Affected the Continental United States: 1851-2006.(Updated from Jarrell et al. 1992 and reflecting official HURDAT reanalysis changes through 1914. Note that from 1915 through 1979, no official wind speed estimates are currently available.)

Year	Month	States Affected and Category by States	Highest Saffir-Simpson U.S. Category	Central Pressure (mb)	Max. Winds (kt)	Name
1851	Jun	TX, C1	1	977	80	-----
1851	Aug	FL, NW3; I-GA, 1	3	960	100	"Great Middle Florida"
1852	Aug	AL, 3; MS, 3; LA, 2; FL, SW2, NW1	3	961	100	"Great Mobile"
1852	Sep	FL, SW1	1	985	70	-----
1852	Oct	FL, NW2; I-GA, 1	2	969	90	"Middle Florida"
1853	Oct *	GA, 1	1	965	70	-----
1854	Jun	TX, S1	1	985	70	-----
1854	Sep	GA, 3; SC, 2; FL, NE1	3	950	100	"Great Carolina"
1854	Sep	TX, C2	2	969	90	"Matagorda"
1855	Sep	LA, 3; MS, 3	3	950	110	"Middle Gulf Shore"
1856	Aug	LA, 4	4	934	130	"Last Island"
1856	Aug	FL, NW2; I-AL, 1; I-GA, 1	2	969	90	"Southeastern States"
1857	Sep &	NC, 1	1	961	80	-----
1858	Sep	NY, 1; CT, 1; RI, 1; MA, 1	1	976	80	"New England"
1859	Sep	AL, 1; FL, NW1	1	985	70	-----
1860	Aug	LA, 3; MS, 3; AL, 2	3	950	110	-----
1860	Sep	LA, 2; MS, 2; AL, 1	2	969	90	-----
1860	Oct	LA, 2	2	969	90	-----
1861	Aug *	FL, SW1	1	970	70	"Key West"
1861	Sep	NC, 1	1	985	70	"Equinoctial"
1861	Nov	NC, 1	1	985	70	"Expedition"
1865	Sep	LA, 2; TX, N1	2	969	90	"Sabine River-Lake Calcasieu"
1865	Oct	FL, SW2; FL, SE1	2	969	90	-----
1866	Jul	TX, C2	2	969	90	-----
1867	Jun	SC, 1	1	985	70	-----
1867	Oct	LA, 2; TX, S1, N1; FL, NW1	2	969	90	"Galveston"
1869	Aug	TX, C2	2	969	90	"Lower Texas Coast"
1869	Sep	LA, 1	1	985	70	-----
1869	Sep	RI, 3; MA, 3; NY, 1; CT, 1	3	963	100	"Eastern New England"
1869	Oct &	ME, 2; MA, 1	2	965	90	"Saxby's Gale"
1870	Jul	AL, 1	1	985	70	"Mobile"
1870	Oct *	FL, SW1, SE1	1	970	70	"Twin Key West (I)"
1870	Oct	FL, SW1	1	977	80	"Twin Key West (II)"
1871	Aug	FL, SE3, NE1, NW1	3	955	100	-----
1871	Aug	FL, SE2, NE1	2	965	90	-----
1871	Sep	FL, NW1, SW1	1	985	70	-----
1873	Sep	FL, NW1	1	985	70	-----
1873	Oct	FL, SW3, SE2, NE1	3	959	100	-----
1874	Sep	FL, NW1; SC, 1; NC, 1	1	985	70	-----
1875	Sep	TX, C3, S2	3	960	100	-----
1876	Sep	NC, 1; VA, 1	1	980	80	-----
1876	Oct	FL, SW2, SE1	2	973	90	-----
1877	Sep	LA, 1; FL, NW1	1	985	70	-----
1877	Oct	FL, NW3; I-GA, 1	3	960	100	-----
1878	Sep	FL, SW2, NE1; SC, 1; GA, 1	2	970	90	-----
1878	Oct	NC, 2; VA, 1; MD, 1; DE, 1; NJ, 1;	2	963	90	-----

		I-PA, 1				
1879	Aug	NC, 3; VA, 2; MA, 1	3	971	100	-----
1879	Aug	TX, N2; LA, 2	2	964	90	-----
1879	Sep	LA, 3	3	950	110	-----
1880	Aug #	TX, S3	3	931	110	-----
1880	Aug	FL, SE2, NE1, NW1	2	972	90	-----
1880	Sep	NC, 1	1	987	70	-----
1880	Oct	FL, NW1	1	985	70	-----
1881	Aug	GA, 2; SC, 1	2	970	90	-----
1881	Sep	NC, 2	2	975	90	-----
1882	Sep	FL, NW3; I-AL, 1	3	949	100	-----
1882	Sep	LA, 2; TX, N1	2	969	90	-----
1882	Oct	FL, NW1	1	985	70	-----
1883	Sep	NC, 2; SC, 1	2	965	90	-----
1885	Aug	SC, 3; NC, 2; GA, 1; FL, NE1	3	953	100	-----
1886	Jun	TX, N2; LA, 2	2	973	85	-----
1886	Jun	FL, NW2; I-GA, 1	2	973	85	-----
1886	Jun	FL, NW2; I-GA, 1	2	973	85	-----
1886	Jul	FL, NW1	1	985	70	-----
1886	Aug	TX, C4	4	925	135	"Indianola"
1886	Sep #	TX, S1, C1	1	973	80	-----
1886	Oct	LA, 3; TX, N2	3	955	105	-----
1887	Jul	FL, NW1; I-AL, 1	1	981	75	-----
1887	Aug *	NC, 1	1	946	65	-----
1887	Sep	TX, S2	2	973	85	-----
1887	Oct	LA, 1	1	981	75	-----
1888	Jun	TX, C1	1	985	70	-----
1888	Aug	FL, SE3, SW1; LA2	3	945	110	-----
1888	Oct	FL, NW2, NE1	2	970	95	-----
1889	Sep	LA, 1	1	985	70	-----
1891	Jul	TX, C1, N1	1	977	80	-----
1891	Aug	FL, SE1	1	985	70	-----
1893	Aug	NY, 1; CT, 1	1	986	75	"Midnight Storm"
1893	Aug	GA, 3; SC, 3; I-NC, 1; FL, NE1	3	954	100	"Sea Islands"
1893	Sep	LA, 2	2	973	85	-----
1893	Oct	LA, 4; MS, 2; AL, 2	4	948	115	"Chenier Caminanda"
1893	Oct	SC, 3; NC, 2; I-VA, 1	3	955	105	-----
1894	Sep	FL, SW2, NE1; SC, 1; VA, 1	2	975	90	-----
1894	Oct	FL, NW3; I-GA, 1; NY, 1; RI, 1; CT, 1	3	955	105	-----
1895	Aug #	TX, S1	1	973	65	-----
1896	Jul	FL, NW2	2	973	85	-----
1896	Sep	RI, 1; MA, 1	1	985	70	-----
1896	Sep	FL, NW3, NE3; GA, 2; SC, 1; I-NC, 1; I-VA, 1	3	960	110	-----
1897	Sep	LA, 1; TX, N1	1	981	75	-----
1898	Aug	FL, NW1	1	985	70	-----
1898	Aug	GA, 1; SC, 1	1	980	75	-----
1898	Oct	GA, 4; FL, NE2	4	938	115	-----
1899	Aug	FL, NW2	2	979	85	-----
1899	Aug	NC, 3	3	945	105	-----
1899	Oct	NC, 2; SC, 2	2	955	95	-----
1900	Sep	TX, N4	4	936	125	"Galveston"
1901	Jul	NC, 1	1	983	70	-----
1901	Aug	LA, 1; MS, 1; AL, 1	1	973	80	-----
1903	Sep	FL, SE1, NW1	1	976	80	-----
1903	Sep	NJ, 1; DE, 1	1	990	70	-----

Year	Month	Location	Cat	Pressure	Wind	Name
1904	Sep	SC, 1	1	985	70	-----
1904	Oct	FL, SE1	1	985	70	-----
1906	Jun	FL, SW1, SE1	1	979	75	-----
1906	Sep	SC, 1; NC, 1	1	977	80	-----
1906	Sep	MS, 2; AL, 2; FL, NW2; LA, 1	2	958	95	-----
1906	Oct	FL, SW3, SE3	3	953	105	-----
1908	Jul	NC, 1	1	985	70	-----
1909	Jun	TX, S2	2	972	85	-----
1909	Jul	TX, N3	3	959	100	"Velasco"
1909	Aug #	TX, S1	1	955	65	-----
1909	Sep	LA, 3; MS, 2	3	952	105	"Grand Isle"
1909	Oct	FL, SW3, SE3	3	957	100	-----
1910	Sep	TX, S2	2	965	95	-----
1910	Oct	FL, SW2	2	955	95	-----
1911	Aug	FL, NW1; AL,1	1	985	70	-----
1911	Aug	SC, 2; GA, 1	2	972	85	-----
1912	Sep	AL, 1; FL, NW1	1	988	65	-----
1912	Oct	TX, S2	2	973	85	-----
1913	Jun	TX, S1	1	988	65	-----
1913	Sep	NC, 1	1	976	75	-----
1913	Oct	SC, 1	1	989	65	-----
1915	Aug	TX, N4	4	945	----	"Galveston"
1915	Sep	FL, NW1; I-GA, 1	1	988	----	-----
1915	Sep	LA, 4	4	931	----	"New Orleans"
1916	Jul	MS, 3; AL, 3	3	948	----	-----
1916	Jul	MA, 1	1	-----	----	-----
1916	Jul	SC, 1	1	980	----	-----
1916	Aug	TX, S3	3	948	----	-----
1916	Oct	AL, 2; FL, NW2	2	972	----	-----
1916	Nov	FL, SW1	1	-----	----	-----
1917	Sep	FL, NW3	3	958	----	-----
1918	Aug	LA, 3	3	955	----	-----
1919	Sep	FL, SW4; TX, S4	4	927	----	-----
1920	Sep	LA, 2	2	975	----	-----
1920	Sep	NC, 1	1	-----	----	-----
1921	Jun	TX, C2	2	979	----	-----
1921	Oct	FL, SW3, NE2	3	952	----	"Tampa Bay"
1923	Oct	LA, 1	1	985	----	-----
1924	Sep	FL, NW1	1	985	----	-----
1924	Oct	FL, SW1	1	980	----	-----
1925	No-De	FL, SW1	1	-----	----	-----
1926	Jul	FL, NE2	2	967	----	-----
1926	Aug	LA, 3	3	955	----	-----
1926	Sep	FL, SE4, SW3, NW3; AL, 3	4	935	----	"Great Miami"
1928	Aug	FL, SE2	2	-----	----	-----
1928	Sep	FL, SE4, NE2; GA, 1; SC, 1	4	929	----	"Lake Okeechobee"
1929	Jun	TX, C1	1	982	----	-----
1929	Sep	FL, SE3, NW2	3	948	----	-----
1932	Aug	TX, N4	4	941	----	"Freeport"
1932	Sep	AL, 1	1	979	----	-----
1933	Aug	TX, S2; FL, SE1	2	975	----	-----
1933	Aug	NC, 2; VA, 2	2	971	----	-----
1933	Sep	TX, S3	3	949	----	-----
1933	Sep	FL, SE3	3	948	----	-----
1933	Sep	NC, 3	3	957	----	-----

Year	Month	Location	Cat	Pressure		Name
1934	Jun	LA, 3	3	962	----	-----
1934	Jul	TX, S2	2	975	----	-----
1935	Sep	FL, SW5, NW2	5	892	----	"Labor Day"
1935	Nov	FL, SE2	2	973	----	-----
1936	Jun	TX, S1	1	987	----	-----
1936	Jul	FL, NW3	3	964	----	-----
1936	Sep	NC, 2	2	-----	----	-----
1938	Aug	LA, 1	1	985	----	-----
1938	Sep	NY, 3; CT, 3; RI, 3; MA, 3	3	946	----	"New England"
1939	Aug	FL, SE1, NW1	1	985	----	-----
1940	Aug	TX, N2; LA, 2	2	972	----	-----
1940	Aug	GA, 2; SC, 2	2	970	----	-----
1941	Sep	TX, N3	3	958	----	-----
1941	Oct	FL, SE2, SW2, NW2	2	975	----	-----
1942	Aug	TX, N1	1	992	----	-----
1942	Aug	TX, C3	3	950	----	-----
1943	Jul	TX, N2	2	969	----	-----
1944	Aug	NC, 1	1	990	----	-----
1944	Sep	NC, 3; VA, 3; NY, 3; CT, 3; RI, 3; MA, 2	3	947	----	-----
1944	Oct	FL, SW3, NE2	3	962	----	-----
1945	Jun	FL, NW1	1	985	----	-----
1945	Aug	TX, C2	2	967	----	-----
1945	Sep	FL, SE3	3	951	----	-----
1946	Oct	FL, SW1	1	980	----	-----
1947	Aug	TX, N1	1	992	----	-----
1947	Sep	FL, SE4, SW2; MS, 3; LA, 3	4	940	----	-----
1947	Oct	GA, 2; SC, 2; FL, SE1	2	974	----	-----
1948	Sep	LA, 1	1	987	----	-----
1948	Sep	FL, SW3, SE2	3	963	----	-----
1948	Oct	FL, SE2	2	975	----	-----
1949	Aug *	NC, 1	1	980	----	-----
1949	Aug	FL, SE3	3	954	----	-----
1949	Oct	TX, N2	2	972	----	-----
1950	Aug	AL, 1	1	980	----	Baker
1950	Sep	FL, NW3	3	958	----	Easy
1950	Oct	FL, SE3	3	955	----	King
1952	Aug	SC, 1	1	985	----	Able
1953	Aug	NC, 1	1	987	----	Barbara
1953	Sep	ME, 1	1	-----	----	Carol
1953	Sep	FL, NW1	1	985	----	Florence
1954	Aug	NY, 3; CT, 3; RI, 3; NC, 2	3	960	----	Carol
1954	Sep	MA, 3; ME, 1	3	954	----	Edna
1954	Oct	SC, 4; NC, 4; MD, 2	4	938	----	Hazel
1955	Aug	NC, 3; VA, 1	3	962	----	Connie
1955	Aug	NC, 1	1	987	----	Diane
1955	Sep	NC, 3	3	960	----	Ione
1956	Sep	LA, 2; FL, NW1	2	975	----	Flossy
1957	Jun	TX, N4; LA, 4	4	945	----	Audrey
1958	Sep *	NC, 3	3	946	----	Helene
1959	Jul	SC, 1	1	993	----	Cindy
1959	Jul	TX, N1	1	984	----	Debra
1959	Sep	SC, 3	3	950	----	Gracie
1960	Sep	FL, SW4; NC, 3; NY, 3; FL, NE2, CT, 2; RI, 2; MA, 1; NH, 1; ME, 1	4	930	----	Donna
1960	Sep	MS, 1	1	981	----	Ethel

Year	Month	Location	Cat	Pressure	Wind	Name
1961	Sep	TX, C4	4	931	----	Carla
1963	Sep	TX, N1	1	996	----	Cindy
1964	Aug	FL, SE2	2	968	----	Cleo
1964	Sep	FL, NE2	2	966	----	Dora
1964	Oct	LA, 3	3	950	----	Hilda
1964	Oct	FL, SW2, SE2	2	974	----	Isbell
1965	Sep	FL, SE3; LA, 3	3	948	----	Betsy
1966	Jun	FL, NW2	2	982	----	Alma
1966	Oct	FL, SW1	1	983	----	Inez
1967	Sep	TX, S3	3	950	----	Beulah
1968	Oct	FL, NW2, NE1	2	977	----	Gladys
1969	Aug	LA, 5; MS, 5	5	909	----	Camille
1969	Sep	ME, 1	1	980	----	Gerda
1970	Aug	TX, S3	3	945	----	Celia
1971	Sep	LA, 2	2	978	----	Edith
1971	Sep	TX, C1	1	979	----	Fern
1971	Sep	NC, 1	1	995	----	Ginger
1972	Jun	FL, NW1; NY, 1; CT, 1	1	980	----	Agnes
1974	Sep	LA, 3	3	952	----	Carmen
1975	Sep	FL, NW3; I-AL, 1	3	955	----	Eloise
1976	Aug	NY, 1	1	980	----	Belle
1977	Sep	LA, 1	1	995	----	Babe
1979	Jul	LA, 1	1	986	----	Bob
1979	Sep	FL, SE2, NE2; GA, 2; SC, 2	2	970	----	David
1979	Sep	AL, 3; MS, 3	3	946	----	Frederic
1980	Aug	TX, S3	3	945	100	Allen
1983	Aug	TX, N3	3	962	100	Alicia
1984	Sep *	NC, 3	3	949	100	Diana
1985	Jul	SC, 1	1	1002	65	Bob
1985	Aug	LA, 1	1	987	80	Danny
1985	Sep	AL, 3; MS, 3; FL, NW3	3	959	100	Elena
1985	Sep	NC, 3; NY,3; CT,2; NH,2; ME,1	3	942	90	Gloria
1985	Oct	LA, 1	1	971	75	Juan
1985	Nov	FL, NW2; I-GA, 1	2	967	85	Kate
1986	Jun	TX, N1	1	990	75	Bonnie
1986	Aug	NC, 1	1	990	65	Charley
1987	Oct	FL, SW1	1	993	65	Floyd
1988	Sep	LA, 1	1	984	70	Florence
1989	Aug	TX, N1	1	986	70	Chantal
1989	Sep	SC, 4; I-NC, 1	4	934	120	Hugo
1989	Oct	TX, N1	1	983	75	Jerry
1991	Aug	RI, 2; MA, 2; NY, 2; CT, 2	2	962	90	Bob
1992	Aug	FL, SE5, SW4; LA, 3	5	922	145	Andrew
1993	Aug *	NC, 3	3	960	100	Emily
1995	Aug	FL, NW2, SE1	2	973	85	Erin
1995	Oct	FL, NW3; I-AL, 1	3	942	100	Opal
1996	Jul	NC, 2	2	974	90	Bertha
1996	Sep	NC, 3	3	954	100	Fran
1997	Jul	LA, 1; AL, 1	1	984	70	Danny
1998	Aug	NC, 2	2	964	95	Bonnie
1998	Sep	FL, NW1	1	987	70	Earl
1998	Sep	FL, SW2; MS, 2	2	964	90	Georges
1999	Aug	TX, S3	3	951	100	Bret
1999	Sep	NC, 2	2	956	90	Floyd
1999	Oct	FL, SW1	1	987	70	Irene

Year	Month	States Affected	Cat	Pressure	Winds	Name
2002	Oct	LA, 1	1	963	80	Lili
2003	Jul	TX, C1	1	979	80	Claudette
2003	Sep	NC, 2; VA, 1	2	957	90	Isabel
2004	Aug *	NC, 1	1	972	70	Alex
2004	Aug	FL, SW4, SE1, NE1; SC,1; NC,1	4	941	130	Charley
2004	Aug	SC, 1	1	985	65	Gaston
2004	Sep	FL, SE2, SW1	2	960	90	Frances
2004	Sep	AL, 3; FL, NW3	3	946	105	Ivan
2004	Sep	FL, SE3, SW1, NW1	3	950	105	Jeanne
2005	Jul	LA, 1	1	991	65	Cindy
2005	Jul	FL, NW3; I-AL, 1	3	946	105	Dennis
2005	Aug	FL, SE1, SW1; LA, 3; MS, 3; AL, 1	3	920	110	Katrina
2005	Sep *	NC, 1	1	982	65	Ophelia
2005	Sep	FL, SW1; LA, 3; TX, N2	3	937	100	Rita
2005	Oct	FL, SW3, SE2	3	950	105	Wilma

Notes:

Hurricanes landfalls that do not produce hurricane-force winds along the coast are not included in this list. Two such hurricanes are known: Sep 1888 in MA and May 1908 in NC.

States Affected and Category by States Affected: The impact of the hurricane on individual U.S. states based upon the Saffir-Simpson Hurricane Scale (through the estimate of the maximum sustained surface winds at each state). (TX S-South Texas, TX C-Central Texas, TX N-North Texas, LA-Louisiana, MS-Mississippi, AL-Alabama, FL NW-Northwest Florida, FL SW-Southwest Florida, FL SE-Southeast Florida, FL NE-Northeast Florida, GA-Georgia, SC-South Carolina, NC-North Carolina, VA-Virginia, MD-Maryland, DE-Delaware, NJ-New Jersey, NY-New York, PA-Pennsylvania, CT-Connecticut, RI-Rhode Island, MA-Massachusetts, NH-New Hampshire, ME-Maine. In Texas, south refers to the area from the Mexican border to Corpus Christi; central spans from north of Corpus Christi to Matagorda Bay and north refers to the region from north of Matagorda Bay to the Louisiana border. In Florida, the north-south dividing line is from Cape Canaveral [28.45N] to Tarpon Springs [28.17N]. The dividing line between west-east Florida goes from 82.69W at the north Florida border with Georgia, to Lake Okeechobee and due south along longitude 80.85W.)

Occasionally, a hurricane will cause a hurricane impact (estimated maximum sustained surface winds) in the inland portion of a coastal state but not at the coast of that state. To differentiate these cases versus coastal hurricane impacts, these inland hurricane strikes are denoted with an "I" prefix before the state abbreviation. States that have been so impacted at least once during this time period include Alabama (IAL), Georgia (IGA), North Carolina (INC), Virginia (IVA), and Pennsylvania (IPA). The Florida peninsula, by the nature of its relatively narrow landmass, is all considered as coastal in this database.

Highest U.S. Saffir-Simpson Category: The highest Saffir-Simpson Hurricane Scale impact in the United States based upon estimated maximum sustained surface winds produced at the coast.

Central Pressure: The observed (or analyzed from peripheral pressure measurements) central pressure of the hurricane at landfall.

Maximum Winds: Estimated maximum sustained (1-min) surface (10 m) winds to occur along the U. S. coast. Winds are estimated to the nearest 10 kt for the period of 1851 to 1885 and to the nearest 5 kt for the period of 1886 to date. (1 kt = 1.15 mph.)

* - Indicates that the hurricane center did not make a U.S. landfall (or substantially weakened before making landfall), but did produce the indicated hurricane force winds over land. In this case, central pressure is given for the hurricane's point of closest approach.

& - Indicates that the hurricane center did make a direct landfall, but that the strongest winds likely remained offshore. Thus the winds indicated here are lower than in HURDAT.

- Indicates that the hurricane made landfall over Mexico, but also caused sustained hurricane force surface winds in Texas. The strongest winds at landfall impacted Mexico, while the weaker maximum sustained winds indicated here were conditions estimated to occur in Texas. Indicated central pressure given is that at Mexican landfall.

Additional Note: Because of the sparseness of towns and cities before 1900 in some coastal locations along the United States, the above list is not complete for all states. Before the Gulf of Mexico and Atlantic coasts became settled, hurricanes may have been underestimated in their intensity or missed completely for small-sized systems (i.e., 2004's Hurricane Charley). The following list provides estimated dates when accurate tropical cyclone records began for specified regions of the United States based upon U.S Census reports and other historical analyses. Years in parenthesis indicate possible starting dates for reliable records before the 1850s that may be available with additional research: Texas-south > 1880, Texas-central > 1851, Texas-north > 1860, Louisiana > 1880, Mississippi > 1851, Alabama < 1851 (1830), Florida-northwest > 1880, Florida-southwest > 1900, Florida-southeast > 1900, Florida-northeast > 1880, Georgia < 1851 (1800), South Carolina < 1851 (1760), North Carolina < 1851 (1760), Virginia < 1851 (1700), Maryland   < 1851 (1760), Delaware < 1851 (1700), New Jersey < 1851 (1760), New York < 1851 (1700), Connecticut < 1851 (1660), Rhode Island < 1851 (1760), Massachusetts < 1851 (1660), New Hampshire < 1851 (1660), and Maine < 1851 (1790).

www.ingramcontent.com/pod-product-compliance
Lightning Source LLC
Chambersburg PA
CBHW081802170526
45167CB00008B/3293